T0093289

PHILOSOPHY OF MICROBIOLOGY

Microbes and microbiology are seldom encountered in philosophical accounts of the life sciences. Although microbiology is a well-established science and microbes the basis of life on this planet, neither the organisms nor the science have been seen as philosophically significant. This book will change that. It fills a major gap in the philosophy of biology by examining central philosophical issues in microbiology. Topics are drawn from evolutionary microbiology, microbial ecology and microbial classification. These discussions are aimed at philosophers and scientists who wish to gain insight into the basic philosophical issues of microbiology.

MAUREEN A. O'MALLEY is a Senior Researcher in the Department of Philosophy at the University of Sydney.

PHILOSOPHY OF MICROBIOLOGY

MAUREEN A. O'MALLEY

University of Sydney

CAMBRIDGE
UNIVERSITY PRESS

University Printing House, Cambridge CB2 8BS, United Kingdom

One Liberty Plaza, 20th Floor, New York, NY 10006, USA

477 Williamstown Road, Port Melbourne, VIC 3207, Australia

314-321, 3rd Floor, Plot 3, Splendor Forum, Jasola District Centre, New Delhi - 110025, India

79 Anson Road, #06-04/06, Singapore 079906

Cambridge University Press is part of the University of Cambridge.

It furthers the University's mission by disseminating knowledge in the pursuit of education, learning and research at the highest international levels of excellence.

www.cambridge.org
Information on this title: www.cambridge.org/9781107024250

First published 2014

A catalogue record for this publication is available from the British Library

Library of Congress Cataloging in Publication data
O'Malley, Maureen, A., 1959–
Philosophy of microbiology / Maureen A. O'Malley, University of Sydney.
pages cm
ISBN 978-1-107-02425-0 (hardback) – ISBN 978-1-107-62150-3 (paperback)
1. Microbiology–Philosophy.   I. Title.
QR41.2.O43 2014
579.01–dc23
2014014063

ISBN 978-1-107-02425-0 Hardback
ISBN 978-1-107-62150-3 Paperback

'Microbiology is a beautiful science. Someone should do some philosophical work on it'.   (W. Ford Doolittle, 2003)

Contents

Figures and Tables

The majority of illustrations are new graphics, drawn by Michel Durinx (www.centimedia.org). Permission was granted to use several copyrighted images (see Reference List for full citations).

Acknowledgements

For reading and commenting on various chapters, my grateful thanks to: Ford Doolittle and Elio Schaechter (who read several chapters each), plus Sam Baron, Pierrick Bourrat, David Braddon-Mitchell, Adrian Currie, Michael Duncan, John Dupré, Matthias Grote, Adam Hochman, Andrew Holmes, Gladys Kostyrka, Maria Kronfeldner, Arnon Levy, Alan Love, Staffan Müller-Wille, Tom Richards, Susan Spath and Mike Travisano. Anonymous reviewers, at both the preliminary and later stages of writing, were tremendously helpful in shaping this book. Numerous discussions over the last decade, especially at Dalhousie University, Halifax (Nova Scotia), provided valuable material and philosophical insight into microbiology. Feedback from audiences and co-participants at several ISHPSSB and SANU Philosophy of Biology meetings was crucial to the development of the book. Comments from Departmental colleagues during a seminar I gave at the University of Sydney had a formative influence on the concluding chapter. I also wish to acknowledge the many hundreds of references that I have been unable to cite here for space reasons but which have informed my writing. For illustrations and help with finding copyright-free images, thanks to Michel Durinx (www.centimedia.org). The Australian Research Council and University of Sydney funded most of the research and writing time it took to produce this book.

An introduction to philosophy of microbiology

Philosophy of microbiology might seem like a highly specialized and even esoteric subfield of philosophy of biology. However, there are many good reasons to think that in fact microbes form the basis of all things biological and thus have major contributions to make to philosophy of biology. This chapter, and the book in general, will make that case.

The grounds for a philosophy of microbiology

Microbes are the most important, diverse and ancient life forms on our planet. The science of these organisms, microbiology, is the science of the most significant living entities and their influence on all the rest of life. Many philosophers will need to be persuaded of these claims, and this book will try to do that. Every scientific field has philosophical aspects, from how the objects of study are conceptualized to the ways in which those objects are known, but microbiology's philosophical issues have only just begun to attract sustained attention from philosophers of biology. These philosophical aspects have driven many debates in microbiological research itself. This book will set out some central philosophical issues in microbiology, along with suggestions for how microbiological insight contributes to and even transforms philosophy of biology. I will start by making a case for philosophy of microbiology based on a general appreciation of the microbial world and its significance for all life. If the world we inhabit is indeed a microbial world, then many of the standard philosophical ways in which we conceive biological phenomena and how they are investigated will have to be rethought. Each of the following chapters deals with a particular aspect of that rethinking.

This general project has a number of complications. One of them is that common terms for microscopic life forms are colloquial and contestable. 'Microbe', for instance, is a broad and convenient term that is used to cover a range of microscopic life (see Table I.1). It encompasses all

Table I.1: *Terminology for microbes*

Microbe	A general term used to cover microscopic and usually unicellular life; equivalent in English to microorganism (in contrast to macrobe or macroorganism). It includes visible aggregations of unicellular life, such as biofilms and colonies. It may occasionally include viruses.
Prokaryote	Unicellular life with a flexibly organized intracellular structure that has limited or, more likely, poorly recognized compartmentalization.
Eukaryote	Unicellular and multicellular life forms with many well-known compartmentalized processes in each cell.
Bacteria	One of the two main groups of prokaryotes; also known as eubacteria.
Archaea	One of the two main groups of prokaryotes; also known as archaebacteria.
Protist	Any unicellular eukaryote except for single-cell fungi, such as yeast (usually excluded but not always); multicellular algae are sometimes included.
Virus; bacteriophage	Non-cellular evolving entities able to use cells for reproduction. Viruses use eukaryotic cells; bacteriophage ('phage') use prokaryotic cells. The most inclusive term is still virus, however.

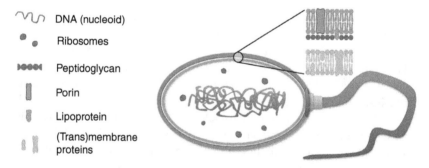

DNA (nucleoid)

Ribosomes

Peptidoglycan

Porin

Lipoprotein

(Trans)membrane proteins

Figure I.1: A prokaryote cell. A schematic diagram of a prokaryote cell, in this case (because of cell wall differences) a Gram-negative bacterium such as *Escherichia coli*.

unicellular life forms (prokaryotes, protists, unicellular fungi and algae) and often includes viruses, even though these entities are not cellular and are rarely considered to be alive in the way that cellular life is. Several of the issues that revolve around formal and informal classification terminologies for microbial life will be discussed in Chapters 2 and 3. A further necessary clarification is that when I discuss the microbial world, I do not refer primarily to the laboratory world of microbiology. Although a vast amount of knowledge has been generated in over a century of laboratory studies, these approaches have obtained limited access to the far greater diversity of

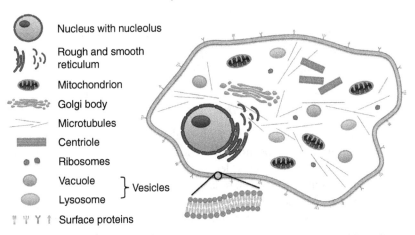

Nucleus with nucleolus

Rough and smooth reticulum

Mitochondrion

Golgi body

Microtubules

Centriole

Ribosomes

Vacuole
Lysosome } Vesicles

Surface proteins

Figure I.2: A eukaryote cell. A schematic diagram of a generalized eukaryote cell (without a cell wall), depicting some specialized compartments and outer membrane structures. Structural features such as flagella, tubulin and actin are not shown, and the cell's size is not proportionate to the prokaryote in Figure I.1.

the uncultured microbial world. This largely unknown world interacts in complex ways with microscopic and macroscopic entities, including the Earth's geochemistry. The book's cover is meant to capture this perspective, and Chapter 5 will develop this theme in some detail. For now, I will simply make the case that the biological world *in general* is microbial. I will do this from the four perspectives of biodiversity, biogeochemistry, evolutionary history and symbiotic collaboration.

The case for a microbial world

The first task in presenting a philosophy of microbiology is to make a case that microbes are of special biological significance. To appreciate their importance, we will consider the quantity, biomass and variety of microorganisms (*biodiversity*). However, biodiversity on its own is not quite enough, even if it is impressive. The next step is to show that the extraordinary metabolic capacities of microorganisms have effects on the planetary processes that sustain all life forms (*biogeochemistry*). These biogeochemical cycles are themselves the products of evolution over the entire history of the Earth, in which microbes have not only themselves evolved but have had major evolutionary impacts on every other evolving life form (*evolutionary history*). This impact largely derives from the multiple capacities microbes have to work with other biological entities (*symbiotic collaboration*). The following chapters will explore each of these four perspectives in more detail. Here they are advanced in order to justify a focus on microbial life, and to support the claim that it needs to be prioritized in any general study of living things.

Biodiversity

Biodiversity is a property of the biological world that is important both ecologically and anthropocentrically. Most biodiversity on Earth is and always has been microbial, regardless of the greater visibility of animals and plants, and despite many differences in how biodiversity is calculated (see Chapter 5). Microbes outnumber all other life forms combined, even though no exhaustive enumeration of them has yet been made and probably will never be. It is unlikely that there is any environment on or around the Earth that is free of microbial life. Microbes can be found in the stratosphere (the atmospheric layer about 15–50 kms above the Earth's surface), in clouds and other condensed water in the troposphere (8–15 kms above the Earth), and as stowaways on materials sent into space by humans. Going in the other direction, there are microbes in the deepest darkest oceans, as well as several kilometres below the Earth's surface. This diversity of habitat is matched in some microbes by an ability to survive for millions of years in a dormant spore state (Lennon and Jones 2011). Even the most determined human efforts to render specific environments totally free of microbes invariably fail, due to individual and collective microbial strategies for endurance and dispersal (Kashefi and Lovley 2003).

There are an estimated 4–6 x 10^{30} prokaryote cells on the planet and about an order of magnitude more of viruses (Whitman et al. 1998). Soil biodiversity is particularly rich, with 10^{16} prokaryotic cells in one tonne of soil – compared to 10^{11} stars in the Milky Way – from which a mere 10 grams may yield as many as 10^7 'species' groups (Curtis and Sloan 2005). To make those numbers more concrete, the number of cells in just one teaspoon of soil exceeds the number of humans currently inhabiting the whole continent of Africa (Editorial 2011). More than 50 per cent of the biomass on the planet is prokaryotic,[1] even though prokaryote cells are on average only one-tenth of the diameter of eukaryote cells and one-thousandth of the

---

[1] This estimate of relative biomass excludes the extracellular material of plants, such as cell walls and structural polymers, and has also been questioned by more recent and much lower estimates of sub-seafloor prokaryotic biomass (Kallmeyer et al. 2012). However, this new estimate assumes very small cell size and low carbon content in prokaryotes in restricted nutritional conditions. Those assumptions are likely to be incorrect and thus the revised estimate too low (Jørgensen 2012).

volume (Whitman et al. 1998). In oceans alone, microbes comprise more than 90 per cent of the total biomass (Sogin et al. 2006), and there are 100 million times more of them – 10^{28} – than there are stars in the entire universe (Editorial 2011). In whichever way the multivalent term of species is conceived, unless solely based on morphology, prokaryotes constitute by far the greatest number of lineages. Even if eukaryotes are considered separately, protists dominate the major groups constituting the eukaryotes (Adl et al. 2012; Chapter 2).

More important than entity counts, biomass and taxonomic diversity, however, is the extraordinary diversity of microbial abilities to generate energy, cope with environmental stresses, adapt quickly to new environments and take advantage of existing ones. Many of these capabilities depend on the metabolic versatility of microbes. Metabolism, the cell-based generation of energy, occurs via reduction-oxidation (redox) couplings that are based on the oxidation of electron donors and reduction of electron acceptors. Different redox pathways have combined in individual microorganisms or groups of them to produce the biogeochemical cycles that maintain life on the planet. Animals and fungi are heterotrophs, using carbon fixed by other organisms, whereas plants are almost all photoautotrophs, which use light energy to fix carbon. Microbes, however, can be heterotrophs, autotrophs or mixotrophs, the last involving the combination of very different metabolic strategies in relation to carbon sources (Madigan et al. 2008; Glossary). Two or more of the highly diverse metabolic strategies found in the microbial world may sometimes be found in the same organism. For example, a single organism may be both an oxygenic and an anoxygenic phototroph, or even a photoheterotroph (photosynthesizers that use organic carbon). Some microbes switch between chemoautotrophy (oxidation of inorganic chemical compounds, including carbon) and organic carbon use (chemoorganotrophy), or between aerobic and anaerobic respiration (Madigan et al. 2008; Glossary).

Numerous microorganisms are extremophiles, which means they can metabolize and reproduce in extreme conditions of heat, cold, acidity, salinity and other seemingly inhospitable environments (Harrison et al. 2013). Microbes in harsh environments, using low-energy reactions, may take thousands of years to generate biomass and divide but they nevertheless manage to survive and reproduce (Hoehler and Jørgensen 2013). Even more microbes enter into sustained metabolic mutualisms with other microorganisms, in which one group of organisms supplies as an end product the metabolic substrate for a differently metabolizing group (Morris et al. 2013). New discoveries of microbial metabolism are being made on a regular basis, because of environment-wide molecular detection strategies and the increasing scrutiny of previously unexplored environments and niches (see Chapter 5). Some of the most unusual metabolic discoveries made in the last two decades were predicted to exist primarily because of thermodynamic possibility and the assumption that microbes will always find a way to exploit potential energy gains (Kuenen 2008).

Although ecologists, policy makers and philosophers are not always in agreement about how to define and measure biodiversity (Faith 2007), they do agree that some level of diversity of life forms is important, and that efforts should be made to preserve known life and its habitats. Microorganisms are intrinsic to the maintenance of plant, animal and fungal biodiversity in ways that will be outlined below. The value we attach to macroorganismal diversity relies largely on microorganismal biodiversity, but despite this relationship it is very rare to hear much said about microbial conservation (see the concluding chapter).

Biogeochemistry

The functional diversity of microbes means that these organisms permeate all life. The global chemistry of life is based on and regulated by microbial metabolisms interacting with the Earth's geochemistry (Falkowski et al. 2008; Dietrich et al. 2006; see Chapter 5). Multitaxon groups of microbes deploying diverse and distinct metabolic pathways bring about most of the biogeochemical transformations necessary for life. The genetic bases of these pathways can be transferred horizontally between evolutionarily distant organisms. The interconnected carbon, oxygen and nitrogen cycles provide many of the major elements essential for life on earth, and microbes are deeply implicated in every phase of these cycles.

It is now a well-established fact that ancestral cyanobacteria were largely responsible for the Great Oxidation Event that occurred around 2.4 billion years ago (Canfield 2005). Chapter 1 will discuss this event and its importance for understanding major evolutionary transitions. Although today plants produce about 50 per cent of the oxygen in our atmosphere – in dependence on the cyanobacteria they captured as endosymbionts a billion years ago – this oxygen is all used up in terrestrial respiration and decay. The maintenance of our oxic (oxygenated) atmosphere is due to marine microbes contributing a net gain of oxygen, because of the way in which they decompose anaerobically in ocean sediments (Kasting and Siefert 2002). Photosynthesis by cyanobacteria in the oceans produces enormous amounts of organic carbon too, thus enabling a wide range of heterotrophic life in marine environments. All secondary producers and consumers, including humans, are further dependent on microbes driving sulphur, iron, phosphorus and manganese cycles (Kolber 2007).

Running the nitrogen cycle is a key biogeochemical role performed by large numbers of prokaryotes with diverse metabolisms. Only bacteria and archaea can accomplish nitrogen fixing, which is the metabolically expensive conversion of unreactive nitrogen gas into more reactive nitrogen compounds. Cyanobacteria fix the majority of marine nitrogen through a variety of methods (Kasting and Siefert 2002). On terra firma, legumes are well known for their bacterial symbioses, in which *Rhizobium* bacteria in root nodules supply plants with fixed nitrogen. The plants provide organic compounds to the bacteria and remove free oxygen, which damages the bacterial enzyme involved in nitrogen fixing. This symbiotic system will feature in Chapter 4. Other plants absorb the ammonia or nitrate produced by free-living prokaryotes. Nitrification is the oxidation of ammonia in soils and water. Mutualistic consortia of nitrifying bacteria work together in this process, with one group oxidizing ammonia (much of which is produced by the microbially assisted decay of organic matter to inorganic chemicals) to nitrite, and then another group converting nitrite to nitrate. In microbial denitrification processes, nitrate is usually converted anaerobically back to nitrogen gases, which can play a role in global warming. This step brings the nitrogen conversion process full cycle. New microbial contributions to previously unknown nitrification and denitrification processes have recently been discovered, leading to major revisions of nitrogen biogeochemistry (Francis et al. 2007).

Closely entwined with the nitrogen cycle and similarly affected by human activity is the carbon cycle. It too is essential to life on this planet and is microbially driven, albeit with considerable input from plants. Microbes decompose organic material, especially plant material, and mediate most of the carbon returned to the atmosphere. In addition, as the outline of photoautotrophy and the oxygen cycle showed, microbes and plants convert inorganic carbon to the organic carbon that is the basis of all non-autotrophic life. Prokaryotes store between 60–100 per cent of the amount of carbon stored in plants, and ten times more nitrogen and phosphorus (Whitman et al. 1998). Viruses, often neglected in biogeochemistry because of not being metabolizers themselves, are now thought to play major regulatory roles in nitrogen, carbon and other cycles due to viruses bursting open (lysing) prokaryote cells, which releases organic material (Danovaro et al. 2008).

From a biogeochemical point of view, therefore, whichever cycle is examined and whatever metabolism is involved, microbes form the basis of all necessary life processes. Even a reader determined to maintain a strictly anthropocentric perspective would need to acknowledge that we humans would not survive without the environmental conditions provided to us by microbes. Furthermore, we would not have evolved without them.

Evolutionary history

Microbes are the dominant life forms not only in today's world, but also in all past eras of the living Earth. The origins of life are exclusively microbial; life until recently was exclusively microbial; life in the future will most probably be exclusively microbial too. If there is indeed life on other planets in other galaxies, it is most likely to be exclusively microbial. Stephen Jay Gould (1941–2002), despite his palaeontological training, observed that the Earth always had been and always would be in the 'Age of Bacteria' (1994). However, the implications of this observation go beyond microbes themselves. The existence of microbial life has been the essential basis for the generation of all other life forms. Eukaryotic life began with unicellular eukaryotes; multicellular life began as a variety of organizations of communal microbes. None of the various scenarios for the origins of multicellularity and important subsequent transitions, such as those to metazoan body plans and flowering plants, happened in isolation from microbes.

Although the earliest dates for the emergence of microbial life are not fully agreed upon, almost all estimates, based on fossil and geochemical evidence, date prokaryote origins somewhere between 3.8 and 3.5 billion years ago. A second extraordinary event in the history of life on this planet was the emergence of oxygenic cyanobacteria 2.7 billion years ago (at the latest), and their gradual conversion of the Earth's atmosphere from anoxic to oxic. The first eukaryotic microorganisms probably appeared about 1.5 billion years ago but it took a further billion years (roughly 600 million years ago) for multicellular eukaryotes to make an appearance in the fossil record (Falkowski 2006; Knoll et al. 2006). Shortly afterwards, a mere 530 million years ago, the Cambrian explosion of modern multicellular metazoan body forms occurred.[2]

An appreciation of this timeline (see Figure I.3) and its utter domination by microbial evolution means that even if the evolution of life is thought of as a stepwise series of major transitions in complexity, and the most

---

[2] These are very rough dates and are much debated. Not only are there unresolved issues about fossils, but there is also considerable conflict between fossil and molecular evidence.

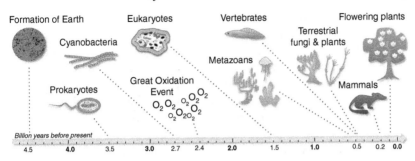

Figure I.3: The evolutionary importance of microbes, with approximate dates.

impressive biological achievements restricted to those that occurred in the last 500 million years, microbial evolution has to be considered in order to contextualize more recent evolutionary developments. Microbes, including viruses, are usefully understood as the engines of evolutionary change, and the several senses in which this is meant will be explored in the following chapters. Evolutionary biologist Theodosius Dobzhansky (1900–1975) is well known for his truism about biology only making sense in the light of evolution (1973). This claim may need microbiological qualification: *all biology and its evolution should be considered in light of microbial evolution.*

Symbiotic collaboration

Multicellular dependence on unicellularity goes beyond a linear relationship in time, and into the present-day constitution of all organisms. A multiplicity of symbioses with microorganisms operates at every level of life. Symbiotic relationships include endosymbioses (within cells), ecto- or episymbioses (on the outside of cells), obligate and facultative dependencies, and mutualist, commensalist and parasitic interactions (Moya et al. 2008). Every symbiosis probably involves microbes, even when the focal symbionts are macrobial. Many mutualistic symbioses are closely integrated and result in the coevolution of the biological entities involved (Herre et al. 1999). I will use the term *collaboration* to describe the flexible symbiotic relationships that have both opt-in and opt-out possibilities, and which involve fluid functional rather than fixed taxonomic relationships (Dupré and O'Malley 2009; see Chapters 4 and 5). These interactions may be reciprocally beneficial in certain conditions; they can endure for millions of years. In others, there may be neutral or negative outcomes for one or more of the participants, but the arrangement nevertheless persists over evolutionary time. More generally, collaboration of various sorts is inescapable amongst living entities and it occurs dynamically at every level of life.

Eukaryotes, although often perceived as instances of autonomous complexity, are defined by their endosymbiont organelles: mitochondria and – in the case of plants, algae and some protists – plastids (see Glossary). The most common of the latter are chloroplasts, the photosynthesizing unit in eukaryotic cells. There are no eukaryotes without mitochondria, and (within eukaryotes) no plants without chloroplasts. Some of these organelles may have reduced or altered functions, such as the divergent mitochondria called mitosomes and hydrogenosomes (Embley and Martin 2006). Some parasitic plants may have greatly reduced chloroplast genomes and no photosynthetic function. However, despite large evolutionary modifications to the ancestral mitochondrion and chloroplast, the designation of 'eukaryote' is based upon microbial collaboration of a fundamental kind. Later chapters will consider a variety of evolutionarily persistent symbioses involving viruses, prokaryotes and other organisms. These relationships exhibit complex balances of cooperation and exploitation, and are maintained by diverse inheritance mechanisms. In many cases, although there are theoretical advantages to non-cooperation, collaborative arrangements between microorganisms and multicellular organisms have given the collective the status of an evolutionary unit.

The diversity and persistence of such arrangements do not mean that microbial collaboration is free of competition, or that competitive interactions are not biologically important. They do imply that we need a better understanding of collaborative processes between organisms, and that microbes will always be involved in such partnerships, whatever realm of life is the focus. A microbiological perspective therefore will lead to a better understanding of evolution, ecology and biology in general. All these areas of research have philosophical aspects, and in addition, microbiology directly informs many standard philosophical questions about biological and evolutionary individuality, evolutionary transitions, and the nature of life. The following chapters will elaborate on these themes in some detail.

To sum up this section about the importance of microbes, all four of these perspectives – biodiversity, biogeochemistry, evolution, symbiosis – point toward the conclusion that from the biosphere to the single organism, and from early life until now, this world is microbial through and through. Microbes may be invisible individually, but collectively they constitute the greatest biological forces on the planet. These four perspectives separately and combined make a case for philosophers of biology to think more inclusively about microorganisms.

Some specific philosophical issues in microbiology

Although it might be easy enough to accept that microbes and microbiology are biologically important, the next question is about how *philosophically* important they are. Some philosophers are sceptical:

> I don't immediately see the philosophical significance of microbes ... microbes just personally strike me as incredibly boring critters ... They're not the sort of thing that I yearn to understand, despite their acknowledged biological significance. Lots of things are biologically significant but are not philosophically significant ... They're just too small! (Anonymous philosopher of science, personal communication 2013).

As the following chapters will demonstrate, reflection on microbiology is indeed significant for philosophy, regardless of size issues. In fact, there might be greater philosophical mileage to be gained from microbes than from non-microbial life forms. Bringing philosophy to microbiology, or microbiology to philosophy, allows an in-between zone of dialogue to emerge from which both communities can benefit. Although I have a few things to say about the relationship between philosophy and science in the concluding chapter, my main aim is to open up a variety of topics and issues in microbiology to philosophical scrutiny, and – contra the claim above – to show just how philosophically significant they are.

Chapter 1 will use two case studies to demonstrate how microbiology has rich philosophical issues to explore. The first scientific sketch concerns photosynthesis and its contribution to the major evolutionary transitions, despite not being recognized as such a transition. The second case will focus on magnetotactic bacteria, which are able to move in alignment with the Earth's geomagnetic field, and have featured in philosophy of mind as exemplars of basic representation and intentionality. After challenging adaptationist interpretations of magnetotaxis, I will look at the implications for teleosemantic accounts of this phenomenon. The rest of the chapter will expand on the importance of these philosophical interrogations of microbiology, and discuss what the implications of such insights might be for philosophy of biology and perhaps even microbiology itself.

Chapters 2 and 3 are about microbial categorization schemes and systematics. After a discussion of the terms microbe and macrobe, I will focus on two extremes of the continuum of standard taxonomic categories: at one end, domains, superkingdoms and kingdoms (Chapter 2); at the other, species (Chapter 3). Classification into domains and superkingdoms has been largely based on new knowledge about microbes, both prokaryotic

and eukaryotic. Major recategorizations of basic biology based on microbial cell type and lifestyle have had an impact on all organismal classification. These transformations, along with the more general and highly
contested categories of prokaryote and eukaryote, form a conceptual
debating ground that allows the discussion of numerous philosophical
questions about taxonomic rank, the reality of such groups and classification practice itself.

Chapter 3 will move to the other end of the classification scale and
outline very briefly a history of how species concepts have been deployed
in microbiology. I will focus on contemporary species concepts and show
why these are problematic in their application to microbes and have not
been generally accepted by microbiologists – not even by those who want
a universal species concept rather than a plurality of pragmatic ones.
A major complication arises from lateral gene transfer and intraspecies
genetic recombination, which are of considerable importance for the
representation of relationships between microbial species. It is now clear
that the very idea of a tree of life – as a map of evolutionary relationships
between species – is challenged if not overthrown by the history of mobile
genetic elements in microbial communities. The molecular revolution in
microbiology is both the source of these problems but also the basis of
high hopes for new ways of thinking about microbial biodiversity –
especially in environmentally based classification practice. Discussion of
these various classificatory schemes will give a good indication of how
microbiology is currently reassessing its core conceptual apparatus in light
of new knowledge and new ways of achieving insight into microorganismal life. For philosophers, more is at stake in these revisions than the
species concept: the very concept of biological hierarchy may be undergo-
ing a transformation.

Chapter 4 is framed by a broad discussion of whether there are major
differences between prokaryote and eukaryote evolution (or perhaps
microbial and macrobial evolution). Phenomena in the prokaryote world,
such as 'directed' mutation and hypermutability, lateral gene transfer and
endosymbiosis, are thought to challenge strictly neo-Darwinian ways of
thinking about evolution. Important contributors to the modern synthesis
of evolutionary biology explicitly excluded prokaryotes and sometimes
other microbes. There is ongoing debate within microbiology about
whether a separate account of prokaryote evolution is needed or a broader
Darwinian one. The roles of competition and cooperation in accounts of
evolution have long been disputed within and without microbiology. This
debate leads to questions about units of selection, and whether individual

organisms (let alone genes) are appropriately conceived as the main units of selection, especially given the highly communal and putatively cooperative nature of microbial organization. Overall, microbiology makes philosophers and biologists confront important ontological issues about the adequacy of a focus on single organisms and lineages, and encourages them to explore whether collaborative adaptive and otherwise co-evolving units might in fact be the appropriate focus of evolutionary and other microbiological study.

The remainder of Chapter 4 considers briefly the possibilities of microbial evo-devo, which would involve the study of the interaction between development and evolution in relation to organisms rarely considered in this light by philosophers. The final theme is evolvability, and whether microbes (especially prokaryotes) are more evolvable than other life forms. On their own, as individual lineages, this is a difficult case to make, but when considered in collaboration with other lineage-forming entities, there appear to be some grounds to think of evolvability as a collectively produced phenomenon.

Chapter 5 applies these evolutionary insights to ecological research, which is now a large, rapidly growing field in microbiology. This was not always the situation, and I will contextualize today's microbial ecology within a short history of how the field developed. This sketch will place particular emphasis on the Delft School of microbiology in the early decades of the twentieth century, and how it was transmitted subsequently so that it eventually supplemented and has perhaps supplanted pure culture approaches. The molecular methods that liberated microbial ecology from the laboratory have provided copious data for biodiversity analyses and phylogenetic reconstruction. But these same data have further problematized the definition of species, and brought back notions of communities, ecosystems and 'superorganisms' as relevant units of ecological analysis, and as causal agents in their own rights.

Community function has important implications for how organisms, multicellularity and reproduction are conceived. I will outline some of the important findings about microbial communities in the human gut to illustrate these points, and to show how mechanistic explanations of community-level effects are being sought in microbial ecology. Different modelling strategies are being deployed to do this, and most of them come from large-organism ecology. Relationships between taxa and area, biogeographical distribution, and community assembly rules differ crucially when applied to the microbial world. Some of these distinctions may eventually feed back into how macrobial ecology is understood. Although microbial ecology can be interpreted historically as a reaction against medically oriented pure culture, the tables have turned somewhat now, with eco-logical methods contributing to broad-scope medical microbiology.

Chapter 6 brings the different topics in the book together in a discussion of how microbes have functioned as tools and model systems for other organismal research. In biochemistry and molecular genetics in particular, microbes became early model systems for all basic life processes. This use continued when standard experimental molecular biology made transitions to genomics, systems and synthetic biology. The molecular modelling of microbes has been enthusiastically matched by experimental approaches to evolution and ecology. In those fields, microorganisms have become indispensable to tests of the central assumptions of evolutionary theory and ecological models. Microbes have also been used as the bases of macroevolutionary experiments, in which microbes are exemplars of capacities normally sought in multicellular organisms. Microbial experimental systems in any field of research have also been criticized, however, and I use these criticisms to examine claims about how microorganisms represent macroorganismal phenomena, and the tractability of microbial systems vis-à-vis macrobial ones. The epistemic gains allowed by microbial systems are not just on the side of science: they are available for philosophers, too. The last part of this chapter reflects on the 'unity of life' assumptions underpinning many uses of microbes as models, and draws further conclusions for philosophers about the value of not just including microbes but in beginning philosophical analyses of life science and living systems with the small things. This, I suggest, enables a big-picture philosophy of biology.

All these themes and cases lead into the concluding chapter, which looks at very broad philosophical questions that need microbes or microbiology to provide any kind of answer. The relationship between anthropocentric and microbe-focused perspectives on life become especially clear when looked at in light of microbial conservation, and whether microbial life forms could and should be preserved. The 'what is life?' questions so beloved of metaphysical inquiries into biology can be answered very interestingly when reflecting on new microbiological findings, such as the huge viruses that appear to fall between cellular and acellular biological entities. Some researchers see big viruses as a challenge to conventional criteria demarcating life and non-life. Anyone thinking about the origins of life is obviously going to require a microbial perspective, but likewise, so will any broader account of life that attempts to understand its nature, origins and basic units. This broad perspective will necessarily find its basis in reflections on microbial evolution and ecology. Evolutionary transitions, which come up several times throughout the book, require a different model from the hierarchical ratchets to which most philosophers of biology have paid attention. I will suggest that a less hierarchy-focused perspective goes hand-in-hand with a microbial metabolism-focused account of biological organization over evolutionary time. Overall, microbiology has many implications for how philosophers can and should approach biology, and I make some methodological suggestions to encompass the strategies I have used throughout the book.

There are many more topics that could be included within philosophy of microbiology than I have managed in this book, and much philosophy of science work that could be done on microbiological modelling and its achievements (the topic of a future book). The few areas I have highlighted illustrate what can be gained by taking a philosophical perspective on microbes and microbiology. The main philosophical foci are the collective entities microbes form, and the relationships between microbial and macrobial models of classification, evolution and ecology. I have not attempted to come up with definitive resolutions of the debates I have outlined, or even tried to suggest entirely novel philosophical messages to take home from microbiology. My aim, simply put, is to place microbes and microbiology on the philosophical map, and to urge more of an engagement between the science and its potential philosophy. The rest of this book is designed to facilitate that engagement.

CHAPTER I

Philosophy in microbiology; microbes in philosophy[1]

This chapter aims to address a question that is central to the motivation for reading (and writing) the rest of the book: why is it valuable to examine microbiology philosophically? As an initial response to this question, I will outline two microbiological case studies. One concerns major evolutionary transitions and how these need a new framework when microbial metabolism is taken into account. The other is about magnetotactic bacteria and their interpretation in philosophy of mind. These cases show in general and specific ways how philosophy can interact fruitfully with microbiology.

Philosophy vis-à-vis microbiology: two sketches

The philosophical importance of microbiology needs further argument, and I will provide a basis for this via two case studies. Each will explore certain capacities of microbes and discuss them in relation to how the science has been interpreted philosophically. One will start from a planetary perspective while the other will enter the inner mental world. The first case concerns microorganismal phototrophy or 'light eating' in the form of oxygen-generating photosynthesis. I will discuss the biological production of oxygen as a major transition in evolutionary history and consider whether its exclusion from the main account of evolutionary transitions is justified. The second case is focused on a microorganism that has had a curious amount of attention in philosophy: the magnetotactic bacterium. Long an example for teleosemantics, which is the philosophical study of how mental content can be explained naturalistically, further analysis of this organism challenges its exemplar status and the standard adaptationist interpretations of its magnetotactic capacities. Both cases give some insight into the ways in which microbes can be useful to philosophy.

---

[1] The title was suggested by Soraya de Chadarevian for the whole book.

Planetary transformation and evolutionary transitions

The overarching question of all biology might be 'why is life like it is'? To answer this question in relation to the life forms that exist today, it is necessary to think about the different turning points in evolutionary history that have enabled major shifts in biological capacities. Photosynthesis and the oxygenation of the planet comprise one of those shifts. In the introductory chapter, I noted that phototrophy refers to the biological use of light to generate energy. A sophisticated means of doing so is photosynthesis, which only in some photosynthesizing organisms produces oxygen. While many eukaryotic oxygenic photosynthesizers are known today (for example, plants, algae, various protists), they all trace their photosynthetic capabilities to ancestral cyanobacteria. Oxygen is the waste product produced by these photosynthesizers, which use 100 or more proteins plus chlorophyll (a green pigment protein) to capture light energy to oxidize water and fix carbon. Cyanobacteria today are numerous and very diverse, with many variations in genome size, physiological capacities, and proclivities for multicellularity and symbiosis (Bryant and Frigaard 2006). They can live in an extraordinary variety of environments, including extremely hot, cold, dry and low nutrient habitats (Scanlan 2001). Although oxygenic cyanobacteria had evolved around 2.7 billion years ago (and possibly as early as 3 billion years ago), oxygen levels only began to rise noticeably around 2.4 billion years ago due to interactions between biogenic oxygen and geochemical events (Sessions et al. 2009; Buick 2008). Oxygen reached its current levels in the atmosphere and aquatic environments only half a billion years ago, after eukaryotes had acquired cyanobacterial cells (see below) and began to dominate the oceans and later, land.

Although there are still important unknown intracellular and biogeo-chemical aspects of oxygenic photosynthesis, especially to do with the mechanisms of water decomposition and the planetary regulation of oxygen levels (Falkowski and Isozaki 2008), it is widely accepted now that classic oxygenic photosynthesis arose fairly late in the flurry of metabolic innovations that characterized the early era of life.[2] This form of photosynthesis was built by combining the two separate photosystems (reaction

---

[2] Earlier interpretations of very ancient microfossils as oxygenic photosynthesizers are not supported by further examination, and the fossils might not even be of organisms (Blankenship et al. 2007; Buick 2008). However, as noted above, cyanobacteria could have been producing oxygen as early as 3 billion years ago.

centres) of anoxygenic photosynthesis, which had been around since nearly the beginning of life on Earth (Allen and Martin 2007; Blankenship et al. 2007). These much earlier photosynthesizers, the descendants of which fall into several major evolutionary groups, each have only one of the systems used in oxygenic photosynthesis – either photosystem I or II (Falkowski and Isozaki 2008; Bryant and Frigaard 2006). Gene transfers[3] between different lineages of these anoxygenic photosynthesizers is one explanation of how these two photosystems combined in cyanobacteria (Knoll 2003). Another explanation is that there is a 'missing link' organism that had both systems. It switched back-and-forth between systems until one sunny day in the Neoarchaean era (2.8 to 2.5 billion years ago), the switch genes mutated, stopped working, and the combined oxygenic system evolved (Allen and Martin 2007). This 'proto-cyanobacterium' could still exist today, despite no signs of it anywhere. A further possibility is that all photosynthesizers originally had both systems for anoxygenic photosynthesis, but except for cyanobacteria, other lineages lost one photosystem or the other (Olson and Blankenship 2004). Presumably, these dual-system proto-cyanobacteria were anoxygenic photosynthesizers until they worked out how to use water as an electron source. The pigment in anoxygenic photosynthesis is bacteriochlorophyll, which is also thought to be ancestral to chlorophyll (Knoll 2003), but the evolutionary routes of how one evolved into the other do not map neatly on to any account of photosystem evolution (Blankenship 2010). Other elements of the photosynthesis machinery (for example, light-harvesting antennae, the electron transport chain) seem to have yet other origin stories.

Anoxygenic photosynthesis had already divided the early ocean into chemical strata, due to photosynthesis occurring in the photic zone (the depth to which light reaches) and substrates in that region being oxidized (Canfield et al. 2006). However, nearly universal anaerobic respiration eclipsed occasional tolerances of the oxygen produced by non-biological and rare biological processes. Oxygenic photosynthesis, in combination with carbon burial by geological and biological forces, upset all these earlier redox balances by providing a powerful oxidant that changed chemical environments and evolutionary possibilities. Its toxicity meant that early cyanobacteria must have rapidly evolved mechanisms to cope with the poison they produced. Metabolic networks in cells show signs of dramatic

---

[3] Transfer of genes between organisms (rather than inheritance from a parent) is a process known to occur in prokaryotes and other organisms. It is called lateral gene transfer and will be discussed in detail in Chapters 3 and 4.

expansion after the rise of oxygen, with hundreds of new reactions and metabolic pathways coming into existence (Raymond and Segrè 2006). Metabolic productivity, especially primary productivity (autotrophy), is considerably higher in the presence of oxygen than in anaerobic regimes, but quite how much higher is still debated (Canfield et al. 2006).

It took another major evolutionary event to spread oxygenic photosynthesis even further. Following the unique engulfment of a bacterium by an archaeon to produce a hybrid cell (the eukaryote) with its clade-distinguishing mitochondrion (the remnant of the engulfed bacterial cell), at least one of these eukaryotes continued its acquisitional evolution and enclosed an ancestral cyanobacterium. This event occurred around 1.5 billion years ago, and it proved so evolutionarily successful and physiologically popular that it happened a few times more. However, these additional acquisitions not a primary plastid endosymbiosis, which is the event in which plants and algae acquired the photosynthesizing organelle called the chloroplast (see Glossary; Chapter 4), but were secondary and tertiary endosymbioses. Tracing all these events is an intricate phylogenetic puzzle-solving exercise that demonstrates the power of molecular phylogenetic methods to illuminate ancient and major evolutionary events (Howe et al. 2008; Reyes-Prieto et al. 2007). These endosymbiotic events should also be of tremendous relevance to the standard model of evolutionary transitions, which focuses on the formation of new units of selection (see below). Each event of plastid endosymbiosis describes an event in which two organisms became one.

Eukaryotic photosynthesizers in water and on land thus all owe their plastids to cyanobacteria, which themselves continue to excrete oxygen and maintain very particular conditions in atmospheric and aquatic environments. In the oceans, phytoplankton – the collective label for unicellular free-living photosynthesizers, eukaryotic and prokaryotic – are tiny organisms that in terms of global biomass barely register. Plants form the bulk of photosynthesizing biomass. Nevertheless, phytoplankton are responsible for almost half of the carbon fixing that occurs anywhere on Earth, and they are the dominant contributors to biological oxygen production (Falkowski et al. 2008; Kasting and Siefert 2002). Cyanobacteria and algae are also found in numerous endosymbioses with animals, especially sponges, jellyfish, sea anemones and corals (Venn et al. 2008). In addition, unicellular photosynthesizers form a great variety of symbioses that do not occur within cells, particularly with plants (whose evolutionary fortunes were made by endosymbiotic cyanobacteria in the first place) and fungi (in the form of lichens). Even though eukaryotic photosynthesizers acquired their photosynthesizing capabilities from prokaryotes, it would still appear inarguable that the first plastid endosymbiosis, and the origin of oxygenic photosynthesis itself, need recognizing in any schema purporting to capture major evolutionary transitions.[4]

The oxygen and photosynthesis story tells us a great deal about why life is the way it is now. Indeed, '85 % of Earth's and life's evolution has merely been an adjustment to more oxygen' (Buick 2008: 2741). This surely by any standards qualifies the oxygenation of the Earth as a major event. However, this combination of transformations is not part of the standard model of evolutionary transitions formulated by John Maynard Smith (1920–2004) and Eörs Szathmáry (1997; Szathmáry and Maynard Smith 1995). Instead, their model foregrounds innovations in replication that had large impacts on biological organization. 'The transitions must be explained in terms of immediate selective advantage to individual replicators', they argue (Szathmáry and Maynard Smith 1995: 227). Microbes figure in these transitions when they exhibit genetic reorganization with morphological consequences (that is, as the original prokaryotic cell, and then as unicellular eukaryotes[5]). However, half of the eight transitions are concerned with features of multicellularity (obligate sexuality, cellular differentiation, social groups and human communication).

Philosophers have devoted considerable attention to this model's account of novel units of selection, both to add to the model and criticize it (e.g., Calcott and Sterelny 2011). Many efforts have been made to find unity in this hierarchy of biological organization and its causal grounding in informational transmission. However, recent criticisms have argued that 'information' is used inconsistently across the model, and transitions to new levels in the hierarchy are qualitatively different. Daniel McShea and Carl Simpson ponder over these problems:

> We cannot find any theoretical unity in the Maynard Smith and Szathmáry history of major transitions. The list needs revision . . . and not in terms of hierarchical complexity. One could argue that theoretical unity is not their goal . . . But then what is the point of collecting these transitions together? Can it be just their importance as milestones on the road to us? (McShea and Simpson 2011: 31).

---

[4] Russell Powell and I are making that argument in detail in a forthcoming paper.
[5] Oddly, Maynard Smith and Szathmáry subscribe to the protoeukaryote hypothesis, in which eukaryotic cellular organization came into existence *prior* to mitochondrial acquisition (see O'Malley 2010b for a synopsis). This is odd because the 'new unit of selection' is then based on mutational changes in an existing lineage rather than a single unit being created out of two.

There are two closely connected problems here: the list of transitions is disunified, and items have been selected for the roles they play in generating complexity, at the top of which is human consciousness and language. More specifically, important events that fit the 'new units of selection' theme are not included (the primary, secondary and tertiary endosymbioses of the plastid), and planet-wide biological transformations are missing. The oxygen story is just one of the latter.

The Great Oxidation Event does not have any status in Maynard Smith and Szathmáry's classic work,[6] despite geobiologists arguing that,

> from an ecological perspective [oxygenic photosynthesis] might well be regarded as the central event in the history of life, because it liberated biology from hydrothermal vents and other [restricted] environments. . . allowing organisms to spread across the planet (Knoll 2003: 4–5).

From another perspective, of course, the rise of oxygen destroyed an intricate 'electron market place' of donors and acceptors that had evolved in the absence of significant amounts of oxygen (Falkowski 2006: 1724). Nevertheless, there was no looking back to this earlier anoxic Garden of Eden. Anaerobic life had the options of adapting to cope with oxygen by taking up aerobic lifestyles or devising oxygen tolerance mechanisms, managing to avoid oxygen by finding deeply anoxic niches, or becoming extinct (Hohmann-Marriott and Blankenship 2011). As oxygen levels gradually increased over the next billion years or so, this environmental transition very probably enabled the earliest multicellular forms of life, amongst which were filament-forming cyanobacteria. These organisms are very important evolutionarily not just because of their photosynthesizing capacities, but also because they evolved very early many of the hallmarks of multicellularity, including cellular adhesion, intercellular communication, metabolite sharing and division of labour (Schirrmeister et al. 2013).

Although metabolism is not the only thing microbes do, it is of particular importance for a general understanding of evolution and biology. This importance is not just because of the metabolic diversity and sophistication that characterizes the microbial world, but because microbial metabolic capacities influence life on Earth today and illuminate its ancient evolutionary history. When the evolution of photosynthesis and its consequences are considered – hand-in-hand with microbial propensities to sequester carbon in anoxic environments and thus keep the planet in a positive oxygen balance – and especially when the major metabolic

---

[6] Szathmáry is now trying to remedy this (Szathmáry and Fernando 2011).

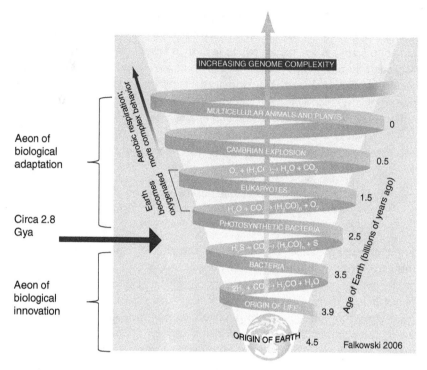

Figure 1.1: Evolutionary transitions defined metabolically. 'Gya' stands for 'billion years ago'. From Falkowski (2006), used with permission from AAAS/*Science*.

pathways of life are taken into account, they suggest that there is probably a much better way to conceive evolutionary transitions.

A promising basis for a new model can be found in the two major aeons of evolution suggested by Paul Falkowski, an oceanographer and marine biogeochemist (2006; Falkowski and Godfrey 2008; Figure 1.1). The first is the 'aeon of biological innovation', during which all existing electron transfer mechanisms and metabolic pathways coevolved. They interact to form the biogeochemical cycles that sustain life today (as discussed in the introductory chapter). The second is an 'aeon of biological adaptation' in which those previously evolved metabolic pathways are shared by gene transfer between organisms, by standard vertical inheritance and speciation, and by the formation of symbiotic collaborations. In many instances, these metabolic capacities are remodelled to fit new morphologies and environmental conditions (Falkowski and Godfrey 2008). The only truly major and unique events in this era are the acquisition of what became the mitochondrion,[7] by either a protoeukaryote or a prokaryote, the latter of which became a eukaryote due to the merger (Lane and Martin 2010; Cavalier-Smith 2002; see O'Malley 2010b for a discussion of this unresolved debate), and the original acquisition of the chloroplast (see above). This claim is not made to diminish other adaptations in this second era – and the metabolic responses to increasing oxygen are many (see Raymond and Segrè 2006) – but to put Maynard Smith and Szathmáry's replication and inheritance mechanisms in better perspective.

This two-phase metabolism-based scheme provides a very broad way in which to understand major evolutionary transitions: as the product of causal interactions between biological and geochemical factors. Although it favours prokaryotes as evolution's most innovative organisms, there are so many of them that this seems more balanced than devoting half the transitions to multicellular life. In addition, the metabolic framework does not lend itself to an apogee interpretation, in which human communication is at the top of the evolutionary transitions (as it is in Szathmáry and Maynard Smith's transitions). Cyanobacteria and other organisms with photosynthesizing machinery should be high on any list of evolution's most remarkable achievements. One question that might be raised in response to this proposal, however, is that many important biological features are not emphasized adequately in this two-era prokaryote- and metabolism-focused schema of major evolutionary transitions. However, all the features that are currently celebrated in Maynard Smith and Szathmáry's transitions can be found in a great variety of microorganisms, as analogues or evolutionary precursors. Social evolution, for example, is found in prokaryotes, and it probably evolved very early in their evolution. Multicellularity is very widespread (see Chapter 4). A two-era metabolic schema can include those events as secondary features that arose in dependence on metabolic capacities.

Another potential argument against the two-phase metabolism scheme comes from philosophers who have suggested that evolutionary transitions are primarily interesting because they can be theoretically unified. Photosynthesis might be seen as a 'possibility-expanding' innovation, but not one that is as theoretically 'profound' as Szathmáry and Maynard Smith's emphasis on new informational mechanisms creating novel biological individuals (Calcott and Sterelny 2011: 4). These philosophical evaluations are

---

[7] For a thorough argument against the selective advantage of mitochondria having anything (or much) to do with oxygen and increased energy yields, see Martin (2007). This essay provides a very useful counterpoint to any suggestion that all the important evolution in the Proterozoic era must have been an adaptation to oxygen.

built on Maynard Smith and Szathmáry's claim that 'There is sufficient formal similarity between the various transitions to hold out the hope that progress in understanding any one of them will help illuminate the others' (Szathmáry and Maynard Smith 1995: 23). The fact that these events do not appear to be unified, despite considerable effort to make them so, suggests that this traditional model may need to be replaced by another framework.

The evolution of metabolism provides a strong basis on which to re-evaluate some of the major episodes of evolution. This framework is at least as intellectually satisfying as talking about 'information transmission' and replicator transformation. The evolution of metabolism in individual cells and biogeochemical cycles is crucial to any explanation of biological individuality and structural innovation. But it is true that metabolism does not foreground the evolution of hierarchy, which is what most philosophers have found interesting in evolutionary transitions. However, as the following chapters show, a microbial-metabolic focus might in fact require the revision of traditional concepts of biological hierarchy and this should surely be of interest to philosophers (see the concluding chapter). Perhaps the fact that metabolism and biogeochemistry are so interwoven has put evolutionary theorists and philosophers off understanding evolutionary transitions metabolically. There could be a concern that because the Great Oxidation Event, for example, and its drivers and consequences are not just about organisms (that is, they include geological and chemical factors), it should not qualify as an evolutionary transition. But if the point is to pick the causal processes relevant to major shifts in evolutionary history, it does not matter what sorts of causes they are. As Szathmáry himself now recognizes, the narrow focus on 'informational' causes without geochemistry was a mistake. He agrees it should now be remedied (Szathmáry and Fernando 2011; Lenton et al. 2004).

A major message from this case study, therefore, is that microorganisms are particularly important to models that encompass large-scale views of evolution, ecology, and life itself. Thinking about the evolution of metabolism and biogeochemistry gives microbes a very prominent role and fills in the big gaps left by large-organism perspectives. Other chapters will reinforce the point that microbes enable a broad, dynamic view of evolution in which the traditional major transitions are radically reconceived along with the very notion of evolvability. The second main message from this case study is about how phenomena often considered the exclusive property of macrobes can be modelled in microbial systems. Several of the transitions Szathmáry and Maynard Smith see as steps toward greater complexity are phenomena such as sociality, multicellularity and communication.

These capacities can be found in many microbes, and there is a large literature and history of experimentation using microbes to model those supposedly macrobial phenomena (see Chapter 6). However, there are reasons against seeing the microbial form as the 'simple' version of the more complex, more philosophically interesting capacity. The second case study helps make that point. It focuses on philosophical discussions of 'mental' representation in which microbes – despite their obvious lack of brains and nervous systems – play a starring role.

Representation, teleosemantics and adaptationism

If any philosopher outside philosophy of biology (and perhaps philosophy of science) knows anything about microbes, it is likely to be about magnetotactic bacteria. Their familiarity has come about because these organisms have been used in philosophy of mind to exemplify claims about representation, intentionality and teleosemantics. The following case study will not only discuss the use and possible misuse of magnetotactic bacteria by philosophers, but will also connect them to questions about the sorts of roles microbes can and should play in philosophical models.

Magnetotactic bacteria are amongst the numerous organisms in the world with internal mineral-based 'compasses'. Other organisms with magnetic-sensing capabilities are mostly animals, such as bees, pigeons and fish, plus an alga. The bacterial compasses are made of nanocrystals of biologically formed minerals (magnetite or, less commonly greigite; very occasionally both) within membrane-bound magnetosomes. These organelles are strung together as chains or sometimes clumps within the bacterial cells. They function as compasses because they orient the bacteria to the Earth's geo-magnetic field. The bacterial flagella, which are the extracellular 'tails' that give motility to many unicellular organisms, follow this passive orientation by actively propelling the bacteria in magnetically defined directions (Bazylinski and Frankel 2004). The bacteria that possess magnetosomes are aquatic microaerophiles (liking very low oxygen concentrations) or anaerobes. Magnetotactic bacteria are ubiquitous, which means they are found globally in aquatic niches. They are diverse, morphologically and physiologically, as well as in regard to magnetosome construction. They do not form a single evolutionary group or clade. This patchy phylogenetic distribution of magnetosome function, plus the location of important mag-netosome genes in genomic 'islands', indicates to phylogeneticists that the organism-to-organism (lateral) transfer of genes involved in magnetotaxis has probably occurred quite frequently (Jogler et al. 2009; Chapter 3).

Microbiologist Richard Blakemore discovered magnetotactic bacteria in 1975.[8] His collaborations with physicist Richard Frankel illuminated the structure of the magnetosome and how it polarizes the cell. Their early work together articulated the sort of question that has guided a great deal of research following the discovery of bacterial magnetotaxis: 'The behavior of the microorganisms, including their movements in response to chemicals or light, provides an adaptive advantage for their survival. Magnetotaxis presumably confers some such survival value, but what is it?' (Blakemore and Frankel 1981: 63).

The authors go on to explain that 'a tendency to migrate downward would be advantageous in that it would help [the microorganisms] avoid the toxic effects of the greater concentration of oxygen in the surface water' (1981: 63). This adaptive hypothesis was confirmed, they believed, by the examination of southern hemisphere magnetotactic bacteria, which did indeed have opposite polarities and thus the same tendency to head downwards (Figure 1.2). The authors then raised the possibilities of the magnetosome serving 'additional purposes', such as a storage site for excess iron or a cleansing device for damaging hydrogen peroxide (Blakemore and Frankel 1981: 65). However, the original hypothesis caught on, and only very occasionally in the literature is any other function (or non-function) of the magnetosome discussed, let alone tested.

A well-known philosophical account of magnetotaxis draws on this standard evolutionary interpretation of its function. Teleosemantics is the label for a project that seeks to explain, in a naturalistic way, mental content, meaning and intentionality. Natural selection becomes the explanation of why representations that do not correspond to the world are eliminated from the evolved capacities of representing organisms. Including organisms other than humans becomes important in this project, because the basic process can be idealized in the model and complex desires, beliefs and inferences set aside. In addition, it is unlikely that representation came into existence only when humans evolved. Although various animals are discussed in the teleosemantic literature, some core teleosemantic arguments use microbes to convey particular points. Magnetotactic bacteria are not the only microorganisms deployed in this philosophical literature,[9] but they are definitely the most discussed.

---

[8] For Blakemore's account of this discovery, see Blakemore (1975). For earlier unpublished studies of magnetotactic bacteria, see Frankel (2009).

[9] Philosopher of mind and cognition, Jerry Fodor (1986), used the eukaryotic microbe, *Paramecium*, in his attempt to refute the possibility of microbes having representational states.

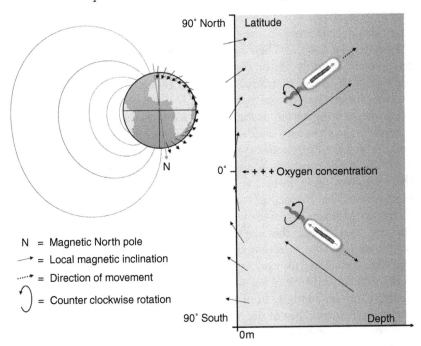

Figure 1.2: Magnetotactic bacteria in northern and southern hemispheres, showing the different polarizations of cells in each hemisphere relative to movement through changing oxygen concentrations in water.

Philosopher Fred Dretske began this discussion of magnetotactic bacterial representation by proposing that although the survival function of the magnetosome could be argued to involve 'representing' the location of low-oxygen niches, since oxicity is merely correlated with magnetic direction, the representing function is therefore of the magnetic field. Even if the bacterium ends up in dangerously oxic waters, the orientation function of the magnetosome is still achieved. This means that misrepresentation does not occur. This is a problem because representation cannot exist without it (Dretske 1986: 29, 32). In her rebuttal of Dretske's point, Ruth Millikan became the philosopher who gave these organisms their enduring philosophical prominence. She uses magnetotactic bacteria to illustrate her thesis that 'an appeal to teleology, to [selected] function, is what is needed to fly a naturalistic theory of content. Moreover, what makes a thing into an inner representation is, near enough, that its function is to represent' (Millikan 1989: 283).

Her take on teleosemantics is that a representation is produced by a mechanism that has the 'proper' function for the organism of producing that representation, which corresponds to the world. Because of performing that function effectively in the past, the mechanism has been selected for (Millikan 1989; 2007).[10] These biological representation devices do not have to represent correctly all the time or even most of the time: they should 'perform their proper functions...just often enough' (Millikan 1989: 289).

Beavers signalling danger by splashing water with their tails, and honeybees performing nectar-locating dances are amongst Millikan's biological examples. She elaborates on Dretske's earlier use of magnetotactic bacteria and the way their magnetosomes appear to guide them into anoxic water. Whether in the northern or southern hemisphere, magnetosomes orient the organisms away from the oxygenated water surface, although the magnetosome has to polarize the organism differently in each hemisphere. However, Millikan disposes of Dretske's 'misrepresentation' problem by claiming that 'What the magnetosome represents then is univocal; it represents only the direction of the oxygen-free water' (Millikan 1989: 291). Described in this way, says Millikan, the magnetosome example bodes well for a theory of mental content: the magnetotactic bacteria do not have to infer anything from the stimulus to the thing represented, and the representation is not of the magnetic field but of the survival-favouring low-oxygen environment. The bacteria can be in the wrong conditions (such as when artificial magnets are applied to the oxygenated water surface), which thus cause the representation to fail. The ability to misrepresent is therefore crucial support for the theory (Shea 2006).

My aim here is not to assess whether teleosemantics is a good way in which to think about mental content (for that, see Neander 2012; Godfrey-Smith 2006). My focus is on the role magnetotactic bacteria are playing in philosophy of mind. Most generally, they function as minimal model systems for representational processes; more specifically, their magnetosome representations are understood as evolutionarily selected. Let me address the more specific issue first. The bacteria are meant to have evolved magnetotaxis in order to orient themselves in relation to the Earth's magnetic poles. Why would they need to do this? The standard answer, which Millikan and Dretske echo, is so that they head for anoxic sediments rather than the oxic surfaces of the aquatic environments in which they live (Figure 1.2). Magnetotaxis seems, however, a very elaborate mechanism by which to orient an organism to murky places: many sediment dwellers do

---

[10] 'Function' in philosophy of biology has two interpretations: one is causal function (what something does for a system) and one is selected function (what something was selected to do). 'Proper function' is the latter.

this perfectly well by more direct means (that is, they work out locally where they are, either in relation to darkness and light, or to low versus high oxygen levels). A correlation between being anaerobic and magnetic field sensing has been interpreted as adaptive causation: organisms head for low oxygen *because of* their magnetosomes.

Some researchers have therefore seen natural design in the structure of the magnetosomes: 'The existence of magnetotactic bacteria is testimony to the power of evolution, as the intracellular magnetic particles reflect the optimization processes of natural selection' (Kopp and Kirshvink 2008: 57). There are at least two ways we might counter this optimal design claim about magnetotaxis: one, that magnetotaxis could be an example of the diverse and not always most efficient or effective ways in which organisms can find solutions to survival and reproduction problems (that is, bad design); two, that surface-sediment orientation is not the primary function of magnetotaxis, which has evolved for other reasons – not necessarily useful for teleosemantics. There is now a range of evidence that questions the simple adaptive hypothesis, and may even put paid to it altogether.

One of the challenges to standard adaptive claims about magnetotactic bacterial orientations – in which magnetic orientation enables oxygen avoidance – is that some northern hemisphere bacteria are oriented to the south and the presence of higher oxygen. Despite this, they survive and proliferate alongside more conventionally oriented bacteria (Simmons et al. 2006). These 'antiparallel' organisms, which move against the geo-magnetic field, appear to be following chemical gradients that would enable different redox reactions. The scientists who discovered this phe-nomenon, in addition to suggesting that current adaptive explanations of magnetotaxis need revision, expressed concerns that adaptive explanations were based too exclusively on evidence gathered in laboratory experiments (Simmons et al. 2006). Laboratory environments often select organisms for capacities they do not exhibit in the wild. Multicellular magnetotactic bacteria (on which more below and in Chapter 4), in a movement known as 'ping-pong' motility, swim very quickly against the magnetic field and then slowly with it (Shapiro et al. 2011). In addition, equatorial magneto-tactic bacteria are oriented horizontally, which is thought to be advanta-geous because at least they are not going further up the water column (Blakemore and Frankel 1981). Horizontal orientation might mean effi-ciencies of escape, dispersal and foraging (Frankel et al. 1981).

Millikan (2007) argues that even if southern hemisphere bacteria end up in the north, with their magnetosome representations failing them, this still does not fail her theory because the general correlation between north and

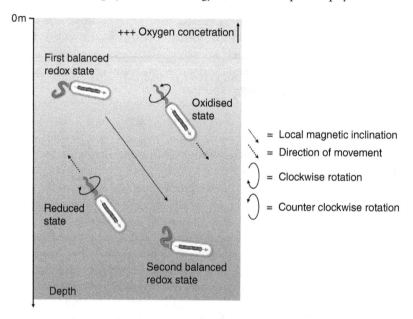

Figure 1.3: A metabolic function for the magnetosome?

less oxygen continues to be faithfully represented. There is additional evidence, however, that complicates the thesis that magnetic orientation represents low oxygen. Many magnetotactic microorganisms are 'aerotactic' too, meaning that they can detect and move in accordance with preferred oxygen concentrations (Bazylinksi and Frankel 2004). A few are even phototactic in response to light (Shapiro et al. 2011). In fact, light exposure makes these particular northern hemisphere bacteria switch back from 'mistaken' south-seeking motility to 'correct' north-seeking. But one way of including aerotaxis in the standard adaptive mechanism is to say that magnetotaxis 'makes aerotaxis more efficient by reducing random excursions and promoting straight-line motility' (Blakemore and Frankel 1981: 65). The south-seeking magnetotaxis in the northern hemisphere can also be explained with an additional redox-seeking motility function that supposedly preserves the standard model of geomagnetic orientation (Zhang et al. 2010; Figure 1.3). Teleosemantics would not have a problem with additional functions supplementing the originally postulated function, and these other functions might have their own representation-producing mechanisms. However, there are bigger problems in the magnetotaxis example than the complications of additional mechanisms and representations.

As already noted, the southern-oriented bacteria in the northern hemisphere seem to be moving in response to chemical gradients, and magnetosomes might be mediating that response (Simmons et al. 2006). An alternative adaptive explanation of magnetotaxis is, therefore, that the magnetosome might be a rechargeable battery that accepts and donates electrons as it encounters different states of oxidation and reduction on the chemical gradient followed by the organism (Kopp and Kirschvink 2008; Komeili 2012; Figure 1.3). It is quite difficult, however, to make redox biochemistry representational. For example, humans respire oxygen, but does this mean that human cells are 'representing' the oxic state of their environments and that this is the 'proper function' of aerobic respiration? This seems unlikely. Another adaptive explanation of magnetosomes is that they function to dispose of reactive oxygen species, such as hydrogen peroxide, which can kill cells (Guo et al. 2012). Eukaryotes have a specialized organelle for this called the peroxisome, which performs other functions too, and its existence is usually explained adaptively. Again, while processing hydrogen peroxide is functional and could well have been selected for, it is very difficult to conceive of it as representation.

Far more problematic for teleosemantics, however, is the existence of non-adaptive explanations of the magnetosome. One suggestion is that magnetosomes are vestigial structures of a now dysfunctional redox reaction (Komeili 2012). A non-adaptive scenario that encompasses this possibility would be that the bacteria need low oxygen to synthesize the magnetite that forms the crystals. In this case, it does not matter what the magnetosomes are *for*; what matters is that the anaerobic environment-seeking behaviour persists. The magnetosomes might just be a side effect of that behaviour, which evolved for other metabolic reasons. One indication this could be right comes from the induction of magnetism in iron storage and redox regulation pathways in genetically manipulated yeast, which are not naturally magnetized (Nishida and Silver 2012). This nice experiment gives tentative support to the 'inadvertent' magnetization hypothesis. Moreover, the cost of producing and maintaining these organelles is possibly not very high, because some bacteria have far more of them than are needed if the magnetic orientation model were true. This redundancy indicates low cost (Spring et al. 1993; Komeili 2012).

How strongly supported are these non-adaptive scenarios? The evidence for them is not particularly strong yet, nor rigorously examined in light of evidence supporting the standard explanation. They are currently more expressions of concern about the adaptationist impulses that have permeated magnetotaxis research. Giving an evolutionary account by explaining a feature with what appears to be merely a plausible narrative has often been derided as telling 'just-so' stories (e.g., Gould 1978). English novelist and poet, Rudyard Kipling (1865–1936), wrote several highly implausible *Just So Stories* (1902). A particularly famous one, depicted on the cover of the first edition of this popular children's book, teased readers with the fiction that the elephant got its trunk from a previous nub of a nose after a marathon tug-of-war with a crocodile in the Limpopo River. A rock-python companion helpfully noted all the advantages of the trunk. 'Just-so' in this story is Kipling's motherese for 'exact explanation' of a particular biological phenomenon, the quest for which is what drives the elephant who originally gets a trunk to the Limpopo in the first place.

The scientific basis for the derision of just-so stories is not because of their implausibility but because of the suspicion of underlying pan-adaptationism. There has been a long debate in biology and the philosophy of biology about adaptationism, and how claims about organismal features being selected for should be tested and not taken for granted. Gould and population geneticist Richard Lewontin are famed for their stimulating anti-adaptationist argument, 'The Spandrels of San Marco' (1979). The adaptationist programme, they suggest, 'regards natural selection as so powerful and the constraints upon it so few that the direct production of adaptation through its operation becomes the primary cause of nearly all organic form, function, and behaviour' (Gould and Lewontin 1979: 584–585). The problem, according to them, is that this singular focus on 'selection for best overall design' does not consider non-adaptive explanations. If they are considered, it is only peripherally.

The adaptive hypotheses about bacterial magnetotaxis are not often claiming 'best overall design' (at least explicitly), but they are using what Gould and Lewontin called 'common styles of argument' in adaptationist research: '(1) If one adaptive argument fails, try another ... (2) If one adaptive argument fails, assume another must exist ... (4) Emphasize immediate utility' (Gould and Lewontin 1979: 586–587).

Even though '*consistency* with natural selection as the sole criterion' (Gould and Lewontin 1979: 588) does not describe how magnetotaxis scientists have viewed the magnetosome, many researchers over the decades have indeed focused on modifying the original adaptive claims about bacterial magnetotaxis. Moreover, although Gould was particularly concerned with by-products that had originated non-adaptively and had then become 'exaptive' for other functions, it is not yet clear whether magnetotaxis can be regarded even as an exaptation because it may not have any selectable function.

But another way to think about magnetotaxis in relation to adaptation is to focus on whether adaptationist claims have been useful. One type of adaptationism is methodological, whereby the phenomenon (for example, magnetotaxis) is investigated with a heuristically adaptationist framework. In philosopher Peter Godfrey-Smith's typology of adaptationism, the methodological form follows the precept: 'The best way for scientists to approach biological systems is to look for features of adaptation and good design. Adaptation is a good "organizing concept" for evolutionary research' (Godfrey-Smith 2001: 337). One of the benefits of such a heuristic is that it can lead to revisionary accounts of evolved function, and this is clearly what has occurred and is still going on in magnetotaxis research. However right or wrong these revisions may be, a lot more has been learned about magnetotaxis in the process – although admittedly, most research has been concerned with non-evolutionary structural and mechanistic accounts of magnetosomes.

A compounding problem of research on magnetotactic bacteria is that not much is known about them experimentally: many of the diverse organisms with magnetosomes even now cannot be cultivated in laboratories. Much is still unknown, therefore, about the causal-role function of magnetosomes, making it very difficult to pinpoint selected function. Nevertheless, teleosemantics attempts to identify the salient causal features of magnetosomes for both proximate and ultimate explanation. The broader problem illustrated by 'proper function' accounts of magnetosomes is that minimalist modelling and adaptationism do not work well together. Minimalist models try to explain general phenomena by abstracting away from the detail and focusing on only the causal factors relevant to the general explanation (Weisberg 2007). However, it takes biological detail to counter simplifying adaptationist tendencies in evolutionary narratives - especially those of optimal design. The minimalism driving the teleosemantic use of magnetotactic bacterial models thus runs up against an overly simplified view of evolutionary function. The difficulties of identifying the appropriate causal factors are exacerbated by the way the magnetotaxis model is being generalized to the target human phenomenon of complex representation. The magnetosome, despite being in a unicellular organism, probably has more complex functional roles than teleosemanticists envisage, both causally and evolutionarily. And even if we could agree these roles are all about 'representing', there is no obvious means by which to move from magnetosome representations to the representations produced and consumed by human brains (Shea 2013).

The teleosemantic perspective on magnetotaxis was motivated in part by hopes that the study of a phenomenon such as representation in a supposedly simple organism could reveal the very basis of representation itself. Microbes in general can be considered useful minimalist model systems for philosophy of biology because of their 'stripped-down' biology, and their ability to stand in for organization that is more complex. A minimalist view of representation seems to have been behind both Dretske's and Millikan's use of these organisms as exemplars, rather than any particular interest in microbes or magnetotaxis. Although there may be grounds to question their analyses of magnetotaxis research, the motivation is one that might be a good rule of thumb for a philosophical modelling strategy: pick an analogous microbial phenomenon to the human one you want to understand naturalistically, and see if this clarifies what the phenomenon is, and what the distinctiveness of the human version of this capacity is. Millikan uses this strategy when she compares microbial and animal representation with human capacities: 'Is it really plausible that bacteria and paramecia, or even birds and bees, have inner representations in the same sense that we do? Am I really prepared to say that these creatures, too, have mental states, that they think? I am not prepared to say that' (1989: 294).

She makes several distinctions between magnetosome representations and how humans represent, and finds that 'these six differences between our representations and those of the bacterium or Fodor's paramecium, ought to be enough amply to secure our superiority, to make us feel comfortably endowed with mind' (1989: 297). However, as discussion of the classic evolutionary transitions model showed, simple-to-complex expectations do not always proceed smoothly. Although the 'human superiority' position may be legitimate in a very restricted sense for philosophy of mind, I doubt that the point of doing philosophy of (micro)biology should be to secure human superiority. In many respects, viewing the microbial world from a general scientific and philosophical perspective encourages feelings of inferiority. Microbes are in numerous aspects functionally complex, and their extensive and plastic capacities mean a lot of work in philosophy of biology should accommodate them without expectations of simplicity, even if the ultimate outcome of doing so is the rejection of particular assumed resemblances between humans and microbes. Using microbes in philosophical models may thus entail the recognition that humans are not at the pinnacle of evolved form and function, and this conclusion is borne out in the previous case study of evolutionary transitions.

In an interesting intersection of these issues of adaptationism and simple-to-complex modelling, Gould himself discussed magnetotactic bacteria in his 1980 book, *The Panda's Thumb*. In that book, he republished a much earlier column he had written in 1979 for the popular science magazine, *New Scientist* (Gould 1979; 1980).[11] Gould marvels over magnetotactic bacteria as 'natural precision designer[s]'. However, although impressed by this bacterial capability, he also observes,

> This proposed function for a bacterial compass is pure speculation at the moment. But if these bacteria use their magnets primarily to swim down (rather than to find each other, or to do Lord knows what, if anything, in their unfamiliar world), then we can make some testable predictions. Members of the same species, living in natural populations adapted to life at the Equator, will probably not make magnets for here a compass needle has no vertical component. In the Southern hemisphere, magnetotactic bacteria should have reversed polarity and swim in the direction of their south-seeking pole (Gould 1979: 447).

Clearly, the first hypothesis has been answered in what Gould might have considered a rather question-begging way (by supplementing the original explanation with an account of why horizontal swimming at the equator is adaptive), and the second hypothesis was borne out as Gould notes in a postscript to the 1979 article in his 1980 reprint.

But perhaps even more curiously, and in a way that takes magnetotactic bacteria out of the simple-to-complex mode of thinking and into major issues of 'representation' in science, Gould was most concerned in his discussion of magnetotaxis with the different worlds of perception that supplemented ordinary human senses and interpretive machinery. Far from wanting us to find the 'essence' of human representation in bacteria (as do Dretske and Millikan), Gould instead wanted to celebrate the glimpse magnetotactic bacteria afford us into the world of non-human perception and representation.

> What an imperceptible lot we are. Surrounded by so much, so fascinating and so real, that we do not see (hear, smell, touch, taste) in nature ... But "parahuman" powers of perception lie all about us in birds, bees, and bacteria. And we can use the manifold and penetrating instruments of science to sense and understand what we do not directly perceive (1979: 447).

There are some lovely parallels between this last sentence and statements by two of the earliest and most famous microscopists and proto-microbiologists (Figure 1.4). Antony van Leeuwenhoek (1632–1723), the first microscopist

---

[11] Gould's 1980 republished version of the 1979 article gains some new introductory paragraphs and a postscript.

Figure 1.4: Early microscopists: Leeuwenhoek and Hooke. The contemporary portrait of Hooke was painted by Rita Greer in 2009.

to publish observations about prokaryotes, exclaimed that microscopes changed the whole human view of the world: 'what if one should tell ... people ... that we have more animals living in the scum on the teeth in a man's mouth, than there are men in a whole kingdom?' (1683, in Dobell 1932: 243).

Robert Hooke (1635–1703), the first microscopist to view microbes (not prokaryotes, but larger protists; see Gest 2004), expressed similar sentiments more grandly:

> by the help of Microscopes, there is nothing so small, as to escape our inquiry; hence there is a new visible World discovered to the understanding ... By this the Earth it self, which lyes so neer us, under our feet, shews quite a new thing to us, and in every *little particle* of its matter; we now behold almost as great a variety of Creatures, as we were able before to reckon up in the whole *Universe* it self (Hooke 1665: Preface – original italicized with emphases in plain text).

Not everyone agreed. Natural philosopher Margaret Cavendish (1623–1673) is particularly well known for objecting to the idea that technologically mediated observations could reveal Nature rather than artefact. The only reliable sensory perception, she argued, is ordinary and unaugmented. Good natural philosophy was for her the product of sense perception and rational contemplation (2001 [1666, 1668]: 53, 9). However, the widespread acceptance that instruments merely enhance the senses is undeniably a major achievement of this early phase of microbiology.[12] Making the invisible or

---

[12] However, see below for Ian Hacking's (1983) discussion of the philosophical antirealism that has been directed toward microscopic entities.

unperceived graspable by the human mind, even when this representation has to be mediated by tools, machines and interpretative agreement, is an epistemological consensus wrought by microscopy within a set of practices that eventually became known as science (Wilson 1995; Ratcliff 2009). As a consequence, the way in which microbes have 'taught' humans new strategies and standards of representation is at least as important as the way in which microbes (allegedly) model our modes of representation.

And finally, the philosophical usefulness of magnetotactic bacteria does not stop with claims about representation, from their perspective or ours. A central theme in philosophy of biology these days is how to conceptualize biological individuals (organisms) and evolutionary individuals (e.g., Godfrey-Smith 2009). I will address this topic from several angles in Chapters 4 and 5, but note here that some magnetotactic bacteria function as multicellular organisms of a very interesting sort. They form tightly organized multicellular clusters that reproduce as smaller multicellular clusters (Abreu et al. 2006; Keim et al. 2004).[13] These organizational properties enable them to be the basis of a discussion for how to do conceptual analysis in relation to biological entities, just as their hypothetical representational capacities did. There are many varieties of multicellularity extant in the world, and no doubt many that have disappeared from the evolutionary record. Only a few of these varieties are easily visible today, and philosophers tend to take animals as the paradigm group for discussions of how multicellularity works over ontogenetic and evolutionary time.

Magnetotactic bacteria that are multicellular do not conform to the animal paradigm, and nor do most of the many microbes that have evolved multicellular forms (O'Malley et al. 2013; see Chapters 4 and 5). In light of their diversity and evolutionary persistence, microbes might be a better starting point than animals for discussions of multicellularity as a general phenomenon (likewise development, reproduction and numerous other capacities that philosophers often model based on animals). Not only are microbes more varied and evolutionarily ancient than animals, but many can be studied as they evolve in the laboratory (see Chapter 6). For these reasons, beginning any philosophical analysis of general biological phenomena with microbes is likely to be valuable, since starting with animals narrows and probably biases conceptions of how multicellularity or other phenomena evolved and can be conceptualized. In following this

---

[13] As noted above, some of these multicellular magnetotactic bacteria seek southern orientations even when they are in the northern hemisphere, and these contrarily oriented organisms are also apparently negatively phototactic (Shapiro et al. 2011).

methodological strategy, teleosemantics can be applauded, caveats about adaptationism and simple-to-complex modelling notwithstanding.

Are microbes and microbiology especially philosophical?

Each of these sketches has something interesting to say philosophically, both to correct the models philosophers have built on the basis of scientific ones, and more constructively. But don't all organisms and every episode of research in biology? Is there something special about microbes and microbiology, such that they should be more valuable objects of philosophical attention than, say, plants and animals? Establishing the biological importance of microbes is obviously not the same as establishing the philosophical importance of microbiology or microbes, as the earlier quote from an unimpressed philosopher of science made clear. The science of microbes is at first glance an improbable candidate for a philosophically intriguing science. Microbiology is often taken to be a technology-driven science that was able to get started and succeed only with the advent of instruments that revealed the microbial world to human observers. And rather than aspiring to make progress towards goals of abstract understanding, microbiology's history has often been seen as one of constant anxiety about pressing practical problems (for example, diseases, food spoilage) and how to rid the world of many microbes.

As discussed above, however, microbiology is important because it deals so directly with the fundamental life forces constitutive of all biological processes. Without an understanding of microbes, there can only be a limited understanding of life. Whether philosophers want to understand evolution, development, immunology or even cognition, microbiological knowledge will clarify and deepen understandings of those processes in even the most 'complex' organism. Major swathes of the history of biological experimentation are either microbiological or have their roots in model microbial systems (see Chapter 6). And even the most abstract and metaphysical lines of inquiry about life, individuality and biological reality will be limitedly and probably uninterestingly investigated if thought is not given to microbes as the basis of life and key to understanding biological organization.

But given that microbiology is such a practical, problem-oriented and technologically driven life science, it might seem that philosophical concerns about the science (as opposed to the biology) are less likely to arise explicitly than they might in, say, neuroscience or conservation biology. It is a curious phenomenon, therefore, that the history of microbiology can be read in toto as a history of philosophical issues. From its earliest instantiations, starting with the Royal Society of London's seventeenth-century forays into microscopy, microbiology has been beset by epistemological concerns about the nature of objects revealed by investigation, and what knowledge about microbes means for biological ontologies. Early microbiology, in the form of the microscopy of microorganisms, raised persistent questions about the nature of scientific evidence, the role of technology in inference and how to assess the validity of claims made about non-visible phenomena. These epistemological problems include specific questions about the status of entities that require technology in order to become observable phenomena. Philosopher Ian Hacking thinks he solves this problem by focusing on experimental intervention as opposed to representational issues. He argues, 'Practice – and I mean in general doing not looking – creates the ability to distinguish between visible artifacts of the preparation or the instrument, and the real structure that is seen with the microscope. This practical ability breeds conviction' (Hacking 1983: 193).

This conviction comes about, he suggests, because of very different types of microscopes converging on the same structures, because of centuries of success in removing aberrations and artefacts of microscopic imaging, and because it is possible to 'interfere' in these structures with a range of physical, chemical and biological techniques (Hacking 1983). In fact, things are a bit less straightforward than Hacking supposes, with numerous developments in imaging techniques relying on suppositions about what should be seen (Breidenmoser et al. 2010). But despite these complications of microscopy, Hacking's main point is still of relevance. Nobody, he concludes, is convinced there are cells – or for our purposes, microbes – due to 'a high powered deductive theory about the cell – there is none – but because of a large number of low level generalizations that enable us to control and create phenomena in the microscope. In short, we learn to move around in the microscopic world' (Hacking 1983: 209). In other words, microbes and other individuated cells teach us how representation works in science, and they did this very early in its modern history.

Microbes do this in a variety of nuanced ways, however, and there are large historical differences in how unicellular organisms have been deemed to represent living things. For example, Charles Darwin (1809–1882), a very competent microscopist as well as an evolutionary theorist, drew on microbiological insights for his theory of natural selection, but made his major impact almost exclusively through zoological and botanical evidence (O'Malley 2009). This is because of the visibility and hence the familiarity

of these life forms. Microbes are just too different and have been conceptualized by many people (not Darwin) as too primitive and too 'other' to
be models for evolved life. Winning a battle over the natural selection of
microbes would not have won the battle regarding animal and especially
human evolution. The modern synthesis of evolutionary biology explicitly
excluded prokaryotes from its early versions (Huxley 1942; Chapter 4), but
even in the 1940s, microbial model systems had managed to supplant
macrobial model systems in molecular genetics (see Chapter 6). Other
epistemological issues of particular relevance today revolve around the
practice of taking laboratory cultures as representative of microbial life,
and DNA sequence as the ultimate phylogenetic proxy for microbial
lineages (discussed in Chapters 3 and 4). Similar issues exist in plant, fungi
and animal biology too, but the very visibility and potential quantifiability
of these macroorganisms – by eye and direct counting techniques – subdue
the force of these epistemological anxieties.

To add to these troubles, almost every aspect of microbiology is conceptually problematic. Its entities fit established ontological categories very
ambivalently, as Chapters 2 and 3 will show in their discussions of
categories such as prokaryote, microbe and organism. The historical division of biology into macro- and microbiology implies two quite different
ontological domains, but the extent of that separation – and whether it is
just a historical contingency or something philosophically grounded – is a
question that the philosophy of biology should probably consider. Evolution, for example, has received extensive philosophical treatment from
philosophers. But this has been the philosophy of animal evolution, with
other eukaryotes largely taken for granted and microorganisms sidelined
(O'Malley 2010a). One of the deepest philosophical quandaries in microbial evolution lies in the question of what interspecies transfer of DNA
between the two largest and most diverse superkingdoms (as well as some
between the Eukarya and the other two superkingdoms) means for a
general understanding of evolutionary lineage, and representations of
evolving lineages in the form of the tree of life. Chapters 3 and 4 will
address this question in some detail, along with the general question of
whether prokaryote evolution is qualitatively different from eukaryote
evolution.

And as already noted above, biological individuality is a popular topic in
philosophy of biology (e.g., Pradeu 2012; Bouchard and Huneman 2013),
and microbiological insights are to an important extent informing this
discussion. This is because microbes illustrate very clearly that there may
be a number of ways in which to carve living things into 'organisms' or

individuals of any sort (Dupré and O'Malley 2009). The proclivity of microbes to form organized multicellular entities, either with related or unrelated microbes, or with multicellular eukaryotes via complex symbiotic relationships, makes the designation of 'organism' somewhat less straightforward than it might seem from a zoological perspective. Moreover, the evolutionary persistence of these arrangements indicates degrees of co-adaptation and the potential of symbiotic units to form units of adaptation and co-evolution (see Chapters 4 and 5). Whether perceived from ecological or evolutionary perspectives, microbes cannot easily be left out of any general discussion of how life organizes itself and what the range of meaningful units might be from any of these scientific perspectives.

There are a variety of other philosophical issues that could be addressed via microbes and microbiology, to do with ethics, aesthetics and the social and political shaping of scientific practice (earlier versions of this chapter included these topics). However, my focus for the rest of the book is on how the nature of microbial organization means philosophers need to rethink some cherished evolutionary and ecological assumptions, and yet at the same time, understand how microbes can be used to model very general biological processes.

Philosophy of biology in light of microbiology

The philosophy of biology is a flourishing subdiscipline of the philosophy of science. Philosophical studies of biology have moved out of the shadow of the philosophy of physics – at least in part due to microbiology, which led the molecular revolution driving most of today's biology (see Chapter 6) – and have become an increasingly popular and philosophically innovative area of research. Today's philosophy of biology encompasses molecular biology, ecology, evolutionary biology, neurobiology and numerous other life sciences. The field prides itself on its up-to-date understandings of and orientation to scientific practice. How, then, has it happened that until recently very little microbiology had come to the attention of the philosophy of biology, and the little that had been noticed was only because microbiological achievements had been interpreted in the light of human or other animal biology?

It may be that philosophers of biology agree with important evolutionary biologists who argue that bacteria can claim only biochemical expertise and occupy only leftover environments. For these biologists, macrobes – particularly metazoans – are much more 'obviously' biologically interesting (e.g., Conway Morris 1998; Mayr 1998). It is certainly true that biologists, such as Ernst Mayr (1904–2005), whose research interests were far removed from microbiology, have influenced the philosophy of biology quite considerably. This influence is due in no small part to Mayr's many efforts to define the philosophy of biology, especially in relation to the philosophy of evolution, and to make this a philosophy of macrobial evolution (O'Malley 2010a; see Chapter 4). But even as the philosophy of biology broadened its interests from predominantly evolutionary concerns, microbiology has not simultaneously become a focus of philosophical inquiry. Historians of biology, however, have been more inclusive and some episodes or topics in microbiology have been the subject of philosophically sophisticated historical inquiry.

One major reason for the philosophical neglect of microbiology may be that for much biology, microbes are merely tools or resources for biological research. Although many molecular and genomic discoveries were made on the basis of microbial cells and their components, the interest in these findings for other biologists has often been in relation to what these findings said about genes, genomes and biological processes in general (see Chapter 6). The philosophy of biology seems to have picked up very strongly on those generalizations without thinking back to the specific model systems in this research. Although it could be argued that an active bias is at work to prevent the philosophical interrogation of microbiology – thereby perpetuating a default preference for visible life forms, morphological complexity and human-like characteristics – it seems likely that something more passive is at play (O'Malley and Dupré 2007). And because philosophical inattention to microbes is in no way definitive of the philosophy of biology, the neglect can be remedied. This book is a partial antidote to an ill-founded and passively reproduced habit, and is part of a new wave of philosophy of biology with a growing appreciation of microorganismal life.

Thinking microbiologically leads, therefore, to several ways forward for the philosophy of biology. The first is the most obvious: a zoocentric perspective provides an exceedingly limited framework in which to do philosophy of biological research, so it should be expanded. Second, as the following chapters will make clear, a microbiological perspective dissolves or at least softens a number of apparently real boundaries between macrobial and microbial, unicellular and multicellular, and dependent and independent life. Each one of these blurrings greatly affects present philosophical conceptualizations of biology, especially those concepts concerned with units in the evolutionary processes of selection and speciation. Third, the highly interactive nature of biological entities, from the smallest to the largest, suggests that a theoretical emphasis on collaboration and interactive dynamics may rival the biological understanding we gain through emphasizing individual autonomy and competitive relationships. And because microbes are involved in all these collaborative interactions, philosophers of biology will add depth and scope to their standard conceptualizations of living phenomena by considering macroorganismal phenomena in the context of microorganismal interactions.

Microbiology in light of philosophy

Microbiologists might react with alarm when told their science is potentially philosophical. This reaction is likely to be a product of equating 'philosophy' with 'speculation', as sometimes happens. What philosophers hope to achieve in their studies of any science may not involve empirical research, but it is likely to involve commitments to rigour, scholarship and evidence-based analysis. The activities that preoccupy philosophers of science are not free-floating cogitations, but the warp and weft of scientific practice: inference, concepts, mental manipulations, justification, evaluation and theoretical structures of all sorts. Philosophers may not be as familiar with the actual science as working scientists, but many efforts are made to collaborate with scientists, gain additional training (to the extent of doing additional PhDs), read copiously and stay up to date scientifically by all means possible. None of this is the same as *doing* science of any kind. But it provides a different and sometimes valuable perspective, and that value may occasionally feed back into the science itself.

Microbiology, as I have tried to show above, is a treasure trove of unexplored philosophical issues that will reward philosophers who pay them attention. Will this attention solve problems in microbiology? Yes and no. I do not think that philosophy of microbiology will devise either a principled account of kingdoms or a universal species concept with equal applicability to all organisms everywhere, even the littlest ones. However, philosophy of microbiology might explain why either of those possibilities is unlikely to happen and why it is not a tragedy for microbiology or biology in general (Chapters 2 and 3). Comparing microbial and macrobial evolution from a philosophical perspective is unlikely to produce a grand unified evolutionary theory derived from first principles, but it will show how different aspects of evolution can be understood in relation to different capacities for variation, inheritance and general evolvability (Chapter 4). Probing the philosophical issues in microbial ecology will not generate exceptionless laws or even provide definitionally tight but methodologically useful concepts of ecological units. However, philosophical oversight might illuminate the ways in which microbial ecology is developing, the gaps in its current conceptual or theoretical machinery, and the justifications for applying macrobial models to microbial phenomena (Chapter 5). Philosophical examination of modelling practices in microbiology will not elevate these models to laws, but is likely to show how it is possible to generalize from microbial phenomena to macrobial (Chapter 6). Philosophical reflections on life in general and the roles microbes play in planetary life are not going to produce one tidy definition of life, but such reflections will show how life can be understood, and why microbes have to be included (concluding chapter). Above all, philosophy of microbiology should be able to demonstrate to biologists the shared epistemological and ontological commitments that connect all forms of biological research, microbial and macrobial. This book will raise only some of these philosophical issues, cast a little light on a few of them and leave many more for further discussion at the interdisciplinary space between philosophy and (micro)biology.

CHAPTER 2

Philosophical debates in high-level microbial classification

This chapter will examine a range of the classificatory practices used to order the microbial world. Some of the categories in hierarchical classification are understood as 'real' and others not. Microbiology's quest for a natural classification system has usually been thought about in regard to species. However, other levels in the 'grand hierarchy of life', such as kingdoms and domains, are just as problematic. I will examine several of these higher-level debates and conclude with their implications for classification in general.

Microbes versus macrobes; pragmatic versus natural classification

This book is about a vaguely defined type of life form: the microorganism. The label 'microbe' or 'microorganism' is used to describe any organism too small to be seen without a microscope. Conventionally, this category includes bacteria, archaea (which also used to be called bacteria), protists and many fungi. Microbiology textbooks usually include viruses and virology within the remit of microbiology (e.g., Madigan et al. 2008), although many microbiologists and virologists do not count viruses as microbes (e.g., van Regenmortel 2011). Although microscopic size is an undeniably important feature of microorganismal life, it is also a very problematic criterion. Many of these supposedly microscopic organisms are quite visible to the naked eye, due to their habit of clustering together in large, sometimes colourful, single-taxon colonies (for example, moulds and filaments) or ubiquitous multitaxon biofilms (slimy accretions of microbes). There are even some bacteria that are visible as single cells. One of these is the spherical sulphur bacterium, *Thiomargarita namibiensis*, which is found in seafloor sediments off the coast of Namibia. It can have a diameter of up to 750 μm (1000 μm = 1mm) and now tops the growing list of giant bacteria, all of which have enormous amounts of DNA due to genome duplication (Angert 2012).

Another common criterion for defining a life form as microbial is that it is unicellular. This characterization generates two further problems. The first is that the preferred lifestyle for most microbes is communal. In Chapters 4 and 5, I will look in more detail at the nature of microbial communality and whether it constitutes anything like multicellularity as standardly conceived. For now, however, the main point is that thinking of microbial life as the life of single cells might not be highly appropriate. Large numbers of microorganisms form a variety of collectives that consist of single or multiple lineages. Many of these collectives exhibit different phases during the population's lifecycle, in which masses of cells function as a cohesive entity with divisions of labour. Although there are many reasons to think of this as a different form of multicellularity from the standard one exhibited by, for example, large animals, there are equally many reasons to think that mere unicellularity is not a defining characteristic of the microorganismal world. Instead, as evolutionary microbiologist Thomas Richards suggests (2013, personal communication), to be a microbe is better understood as 'being able to live and reproduce as a single cell', but not necessarily for a whole lifecycle or for every generation of a population.

The second problem is that some of the entities thrown in with microbes are not usually considered organisms because they are not cellular. These troublesome bits of biology are the multitudinous viruses, which pervade every environment and play important roles in the evolution of 'true' organisms. Because viruses have no metabolic function and require other organisms to replicate, they tend to be placed somewhere in-between living organisms and non-living chemicals (Dupré and O'Malley 2009). Intriguing findings of translation and biosynthesis genes in recently discovered huge viruses mean that some viruses might have more biological capabilities than once thought (Raoult et al. 2004), although it is doubtful the viruses can do much with these genes. But it is undeniable that viruses – as well as other mobile genetic entities – form lineages and are subject to natural selection. They often are used as evolutionary models of 'genuine' organisms, and were invaluable model systems for the development of molecular biology and genomics (Chapter 6). Moreover, viruses interact extensively with prokaryotes and other organisms, to the extent of fundamentally shaping their lives and evolutionary trajectories. Viruses are, therefore, often included in the category of microbes because it is generally accepted that microbiology could not fully understand the microbial world if viruses were to be excluded from it (Rohwer and Thurber 2009; Suttle 2007).

Although 'microbe' is a very common way of talking about microscopic life, it obviously admits many inconsistencies. However, it is also clearly the case that conventional terminological devices are not intended to be anything more than terms of convenience to meet practical needs. Nobody expects deep biological meaning to emerge when such terms are used. This is because there is no assumption at all that the groups of organisms covered by the label of microbe or macrobe form genuinely natural groups (unless they mean prokaryote or eukaryote by these terms, which does occur occasionally – see below). However, the overarching system of taxonomy to which systematists and biologists subscribe much more committedly is meant in some sense to reflect a natural order. Naturalist and taxonomist Carl Linnaeus (1707–1778) bequeathed a hierarchy of ever-more inclusive groups to biological classification, starting with kingdoms and moving in a less inclusive direction to phyla, classes, orders, families, genera and species (de Queiroz 1997).[1] There are no doubt many interesting things that could be said about the categories between kingdoms and species,[2] but my first classificatory focus will be on the upper levels of this hierarchy and how they work inside and outside microbiology. Although classification in biology is usually done according to taxonomic principles and rules (Ereshefsky 2008), throughout this chapter and the next I will be concerned with classificatory practice and what it says to taxonomy.

From kingdoms to domains

Kingdoms of life are of immense importance to any understanding of biology, whether that understanding is strictly scientific or at large in the broader public sphere. From a scientific perspective, kingdoms are levels in the hierarchy of life that are often argued to be non-natural categories: 'not actually things that one discovers, but conceptual entities that systematists create' (Cavalier-Smith 2004: 1260; however, cf. Ereshefsky 1991). But even if they are at least 'partly a matter of tradition and taste' (Valentine 1980: 445), there is still something utterly basic about how the primary divisions of life have been understood in recent decades. Major transformations have occurred in relation to the division of the living world into kingdoms, and those kingdoms in relation to a higher rank of 'domains'. It is debates

---

[1] Phylum and family were subsequent additions to Linnaeas's scheme, which had originally included 'varieties' below species (de Queiroz 1997).
[2] The environmental classification section in Chapter 3 discusses phylum-level classification.

about the microbial world and inferences about its evolutionary past that have driven the major re-classificatory movements involving those categories. I will not attempt to reconceptualize any of these categories more stringently, nor even to give a detailed overview of these historical shifts. My aim is primarily to show the influence of microbial classification on macrobial classification.

When people speak of biological kingdoms now, they mostly refer to a recent five-kingdom division of life. This classificatory level has a long history, but plant ecologist Robert Whittaker (1920–1980) proposed its most accepted sense; microbiologist Lynn Margulis (1938–2011) subsequently modified it. Margulis reorganized the three multicellular kingdoms to make them evolutionarily cohesive (Figure 2.1). Historian of biology Joel Hagen (2012) recounts the history of designating kingdoms, outlining how Whittaker reached his final version of the five kingdoms via the broad classes of data he used to carve up the biological world: cell structure, nutritional mode, ecology and evolution. Although Whittaker's kingdoms depended on a primary division of life into monerans and non-monerans – 'Monera' being the old term for the most 'primitive' unicellular organisms – he gave equal kingdom-level status to the four groups of non-monerans or eukaryotes: plants, animals, fungi and protists (Whittaker 1969; Figure 2.1). Whittaker saw his system, and the four- and three-kingdom versions before his, as the inevitable overthrow of the long-existing categorization of life into two kingdoms of plants and animals. Fungi had only recently been moved out of the plant kingdom and given origins other than algae (Martin 1955; Whittaker 1959), and protists remained an admittedly arbitrary kingdom (see below), consisting of the unicellular eukaryotes and other life forms that are not plants, animals or fungi.

The naturalness of these kingdoms is rarely argued but often assumed. Kingdom designations are major indicators of how various fields conceive basic divisions between different life forms. Fungi are a case in point. They are a useful example because so many of them are unicellular, and some major debates about fungi classification revolve around these unicellular members. Even as late as the 1980s, some phylogeneticists thought the fungi formed an artificial group rather than one with evolutionary cohesion (e.g., Woese et al. 1990). They suggested fungi were paraphyletic rather than monophyletic. Monophyly is a well-accepted principle of phylogenetic classification today, and describes groups (clades) that include the most recent common ancestor and all descendants (see Glossary). A monophyletic group is thus an evolutionarily unified group, and has a special status in natural classification. A paraphyletic group, such as some

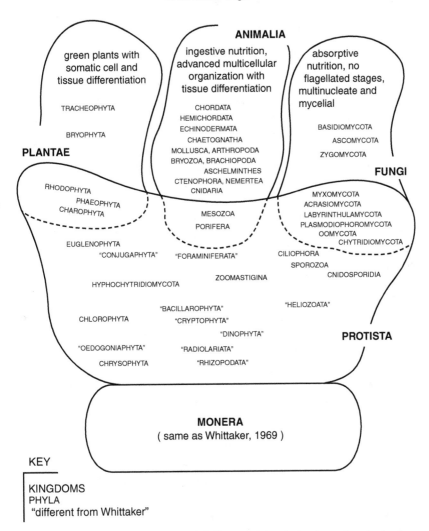

Figure 2.1: Whittaker's five kingdoms, modified by Lynn Margulis (1971) to make animals, plants and fungi into better evolutionary groups. Used with permission from John Wiley and Sons.

people believed fungi to be, includes the most recent common ancestor but not all the descendants (see Glossary). In the 1990s, however, molecular evidence that confirmed the closer relationship of animals to fungi than to plants also consolidated existing notions of the so-called true fungi as a monophyletic group. Today, much larger molecular datasets support the monophyly of fungi (e.g., Ebersberger et al. 2012). Phenotypic data that had for decades contributed in a major way to identifying fungi as a natural group were an abundance of chitin (a structural carbohydrate) in the cell walls of fungi, and the capacity this gives fungi for a distinctive mode of nutritional intake known as osmotrophy. This term refers to how fungi 'digest' organic material outside their cells and then absorb it across the cell wall into the cell; prokaryotes can do it too. Cellular subsystems linked to the chitin, plus osmotrophy, can be regarded as 'unifying characters' of the fungal kingdom (James and Berbee 2011: 96).

However, at the same time that 'true' fungi were being consolidated as a natural group, 'pseudofungi' (for example, unicellular oomycetes, now part of a protist group), were expelled from the kingdom despite having some fungi-like characteristics. The refinement of fungi by discarding uncharacteristic groups has been further complicated by recent findings of a unicellular aquatic fungus, tentatively named cryptomycota or 'hidden fungus'. It has brought both the unifying characteristics and the kingdom-ness of fungi into question (Jones et al. 2011). These organisms, now found in great diversity, go through key lifecycle phases without chitinous cell walls, and they may sometimes be phagotrophic (able to ingest particulate organic matter, including whole cells) rather than osmotrophic. Because of their divergence at the very base of the tree of fungi, cryptomycota appear to be transitional between more obviously protist-like organisms and fungi. They thus blur the kingdom boundaries – if the kingdom of fungi even exists (Jones et al. 2011). Other molecular systematists and mycologists (fungi researchers), however, think these organisms should simply be excluded from fungi because of their lack of chitin (e.g., Cavalier-Smith 2013). But to make matters more complicated, subsequent analyses have detected chitin-synthesizing genes in cryptomycota, and found that the representative taxon, *Rozella*, may have chitin cell walls for a brief phase of its lifecycle (James and Berbee 2011). That same study also suggests that secondary loss of other chitin genes needs to be considered alongside whether the basal placement of the lineage is an artefact of phylogenetic methods.

Debates such as these provide tremendously interesting material for philosophers of classification. There is likely to be even more grist for the philosophical mill as environmental sampling – the way these aberrant fungi were found, rather than via laboratory culturing – continues to lift the lid on cryptic diversity in the microscopic realms of life. Whether such findings concretize or dissolve kingdom boundaries is still to be worked out, but at least for some researchers in the fungi story, burgeoning biodiversity based on molecular analyses is likely to reveal the 'fragility of kingdoms' as a classificatory device (Thomas Richards 2013, personal communication). However, in response to those systematists and philosophers already sceptical of kingdoms, there is an even higher rank that is often discussed as a genuinely natural category.

The demise of the prokaryote?

Whittaker's five kingdoms still exert considerable influence, but he himself observed that it was the division of life into prokaryote and eukaryote cells that was the 'clearest, most effectively discontinuous separation of levels of organization in the living world' (1969: 151). Prokaryotes are usually contrasted to eukaryotes, because the latter possess membrane-bound nuclei and other compartmentalized organelles (Stanier and van Niel 1962; see Figures I.1 and I.2 in the introductory chapter). Prokaryotes generally do not have compartments of the same sort, although there are some very interesting exceptions (see below). Instead, prokaryotes have chromosomes that are not separated from the cytoplasm by a membrane,[3] and this different organizational structure tends to give their possessors different genetic and evolutionary capacities from eukaryotes (see Chapter 4). Reproductive mode is a distinction often made on the basis of what the cells contain. Prokaryotes reproduce mostly by binary fission, indulging only occasionally in genetic recombination in a separate process. They never experience mitosis (the division of nucleated cells), meiosis or syngamy.[4] All eukaryotes have some sort of capacity for these processes, or at least the genes for them (O'Malley et al. 2013). Major advocates of the prokaryote concept thus perceived that 'evolutionary diversification through time has taken place on two distinct levels of cellularity': the prokaryotic kingdom and the eukaryotic (Stanier and van Niel 1962: 33). Although 'prokaryote' and 'eukaryote' have been very

---

[3] Also, prokaryote chromosomes are usually circular while those of eukaryotes are linear. Exceptions are found in the many Actinobacteria with linear chromosomes. Prokaryote genomes were once thought not to have introns until several discoveries of them were made in the 1980s and 90s. Overall, however, prokaryote genomes are richer in genes than eukaryote genomes, despite the presence in the former of introns and other non-coding DNA.

[4] Meiosis refers to the reductive division that produces haploid nuclei from diploid cells. It is accompanied by genetic recombination and the fusion of haploid nuclei. This process normally occurs as part of syngamy, which is the fusion of gametes. Defining the core processes of sexual reproduction in accordance with obligatory sexual reproduction normally excludes prokaryotic genetic recombination. Nevertheless, as we will see in later chapters, 'sex' is often attributed to prokaryotes.

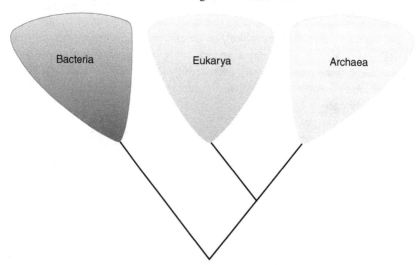

Figure 2.2: Three domains: Archaea, Bacteria, Eukarya. This is the standard representation of the three domains, showing widely accepted evolutionary relationships. Based on Pace (2006).

useful terms for distinguishing two kinds of cellular organization, their biological meaningfulness is much disputed, especially from an evolutionary point of view (Pace 2006; however, cf. Doolittle and Zhaxybayeva 2013).

Just a few decades ago, the attempt to find evolutionary justifications for all classificatory levels brought about a great upheaval to this dualist ontology. Evolutionary microbiologist Carl Woese (1928–2012) proposed a tripartite division of life forms into three 'urkingdoms'. From the vantage point of the old prokaryote-eukaryote classification, he split prokaryotes into the two superkingdoms of life, Bacteria and Archaea. They, along with Eukarya, became the highest groupings in biological classification, and these taxa were eventually labelled 'domains' (Figure 2.2). Woese, who first identified Archaea as an upper level taxon in the 1970s, also coined the label of domain for this taxonomic level. Until then, both prokaryotic domains were thought of jointly as just 'bacteria'. Molecular findings about cell-wall chemistry, metabolic pathways, and transcriptional and translational machinery were the bases of this major distinction between the two cell types (Woese and Fox 1977). At first, the non-bacterial cells were named 'archaebacteria' to distinguish them from the 'true' bacteria, which were called in this taxonomy 'eubacteria'. Although some system-atists still hold to these older labels because they find them more accurate

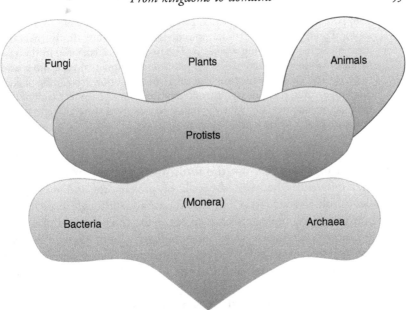

Figure 2.3: Five kingdoms with Monera now split into Bacteria and Archaea. These sorts of depictions, often found at online biology teaching sites, are not really intended to represent evolutionary relatedness, but nevertheless they imply some historical relationships (i.e., that eukaryotes arose from 'Monera', and multicellular organisms from Protists).

and less presumptive of particular evolutionary relationships, Woese and many other evolutionary biologists shifted to the abbreviated and more formal nomenclature of Archaea, Bacteria and 'Eucarya' (Woese et al. 1990).[5] Despite the widespread acceptance of the bacteria-archaea distinction nowadays, most dictionaries and medical references still routinely refer to all bacteria-like organisms (i.e., prokaryotes) as 'bacteria' – perhaps due to a combination of linguistic inertia and the original name for archaea being 'archaebacteria'.

However, there are some much more militantly motivated rejections of the three-domain scheme. The division of the so-called prokaryotes into Bacteria and Archaea was intended to be far more radical than a mere subdivision of Monera (Figure 2.3). Mayr, whose work on species concepts we will encounter in the next chapter, was unable to accept the three-domain division of life as the most fundamental:

---

[5] When 'bacteria' and 'archaea' are used in lower case, this simply means informal use of the formal domain labels. The singular of this informal use of archaea is 'archaeon'.

> I support the dichotomy prokaryotes vs. eukaryotes not owing to a philo-
> sophical preference for a dichotomous division but because this is where the
> great break is in the living world ... The two kinds of bacteria, in the vast
> majority of their characteristics, are exceedingly similar to each other and
> fundamentally so different from the eukaryotes that they have to be ranked
> as a single taxon, the prokaryotes, different from the only other taxon of this
> rank, the eukaryotes. Only a two-empire classification correctly reflects this
> structure of the living world (Mayr 1998: 9723).

Why Mayr or anyone else would want to do this is because even when
multicellular organisms are not the focus, there still appear to be major
biological differences between the cells of prokaryotes and eukaryotes
(Cavalier-Smith 2007; Martin and Koonin 2006). Although most microbiologists and many biologists have accepted the three-domain classification, in a more general sense Mayr's taxonomic viewpoint has won the day.
In many representations, two of Woese's three domains are simply shown
as constituents of the basal group of the old five-kingdom schema, the
Monera (Figure 2.3).

Despite proponents of the three-domain classification scheme recognizing that these three groups had a common evolutionary origin, some
three-domain adherents also believe that the subsequent divergence of
these groups mean that they have 'essentially' different characteristics
(Pace et al. 2012). Early discoveries that established Archaea as a domain
focused on how archaeal systems of transcription and translation, and
DNA replication and repair, were different from bacterial ones. Cell lipids
and membranes, plus certain metabolic features, are also known to be
either unique to Archaea or more similar to eukaryotes. A recently recognized difference is that of the flagellum, the little 'whip' that protrudes
from cells and enables motility. Although it is well known that prokaryote
and eukaryote flagella are vastly different in structure and ancestry, it has
taken longer for attention to be drawn to the very distinct organelle that is
the archaeal flagellum (Jarell and Arbers 2012). All of these differences add
up to the evolutionary interpretation that the supposed group of prokaryotes is not monophyletic, but paraphyletic or worse, polyphyletic.
A polyphyletic group includes only some descendants of a common
ancestor, because that common ancestor is not itself included and does
not determine the group. Such groupings are thus evolutionarily 'mean-
ingless' (Woese et al. 1978; Wheelis et al. 1992; Pace 2006; Glossary).
There are many reasons why the prokaryote-eukaryote distinction is not
disastrous for biology (and these include reasons *for* paraphyletic groups),
but the tendency to reject the distinction on the ground of the three

domains being natural kinds themselves is very problematic (see Doolittle and Zhaxybayeva 2013 for detailed discussion).

Challenges to the three domains

Phylogenetic analyses pose a range of problems for whether the prokaryote-eukaryote distinction or the three domains is more basic to any carving-up of life. Even if a phylogeneticist recognizes that a three-domain division of life need not be essentialist, the very nature of the postulated divergences between the three groups has been challenged, and not just because some of the archaeal distinctions have now been found in bacteria. One view takes issue with the criterion of monophyly itself. Evolutionary microbiologist Barny Whitman (2009: 2003) argues that 'phylogenetic classifications must extend beyond the simple recognition of monophyletic groups' to recognize 'transformative evolutionary events' such as eukaryogenesis. This term refers to the creation of the new cell type, the eukaryote. If the eukaryote cell came about through the fusion of an archaeon and bacterium, as many evolutionary biologists believe it did, phylogeny would have to incorporate somehow a new category of organism that could not be justified monophyletically. In addition to challenges to the phylogenetic principle of monophyly itself, there are monophyly-based challenges to the three-domain model.

Virologists who believe they have found sequence evidence that shows a gigantic form of virus, the Mimivirus, to be a new major clade of life, recently mounted a phylogenetic challenge to the three-domain tree. (Boyer et al. 2010). They claim this group constitutes the 'fourth domain' of life, which raises some major questions not just for classification practice but its broader philosophy. Usually, these large-scale classification schemes are about living things. Viruses – though intensively classified (see 'viral species' in Chapter 3) – are not deemed to be living and thus not part of the tree of life (Moreira and López-García 2009). One tactic for the systematists who argue for the inclusion of viruses is to redefine life altogether or to define a special category of viral life. The 'virocell' concept is part of these efforts, and it proposes that viruses be understood as living when part of a cell, and these cells conceived as 'virus factories' (Raoult and Forterre 2008; Claverie 2006). However, for other systematists, this is simply 'epistemological cheating' (López-García and Moreira 2012: 394) because although viruses are ongoing lineages of biological entities, they do not metabolize or self-replicate – both of which are crucial characteristics of living things. We will

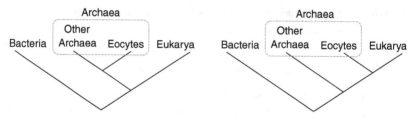

Figure 2.4: The eocyte tree. The left-hand tree shows the standard representation (see Figure 2.2), whereas the right-hand one depicts the eocyte interpretation. Based on Cox et al. (2008).

return to this debate about what life is in the concluding chapter of the book. For now, however, this particular case for a fourth domain is the subject of much criticism due to what are considered to be misinterpretations of crucial DNA and other phylogenetic data (Williams et al. 2011).

The main challenge to the three-domain tree is mounted by the 'eocyte hypothesis', which is now a modernized version of a serious challenge to Woese in the 1980s (Lake et al. 1984). What the old version argued and found evidence to support was that Archaea is not a monophyletic group. Historian of microbial phylogeny Jan Sapp (2009) outlines the history of why this competing hypothesis lost the earlier battle. However, it has been recently revived with the detailed analysis of multiple datasets and improved models of sequence evolution (Cox et al. 2008; Williams et al. 2012). What many of these analyses show is that if Eukarya retains the status of a separate domain, Archaea is indeed a paraphyletic group, and therefore 'unnatural' – an accusation against the category of prokaryote that Woese used to encourage the uptake of his three-domain hypothesis.[6] Archaea is standardly believed to consist of two main phyla, the Crenarchaeota and the Euryarchaeota. In the eocyte hypothesis, the Crenarchaeota (the eocytes) are sister taxa with the eukaryotes, and the other main group of Archaea, the Euryarchaeota, is outside that major clade (Cox et al. 2008; Figure 2.4). To maintain monophyly based on these new data and interpretations, there could be only *two* domains: the Archaea (with the Eukarya within) and Bacteria (Williams et al. 2013). Mayr would have been even more unhappy with this outcome than the three domains. However, the limitations of the evidence, and the difficulties of finding adequate signal and interpreting what signal there is, indicate to the senior investigator of the revised eocyte study, Martin Embley, that a major lesson to draw now is one of epistemic

---

[6] See Doolittle and Zhaxybayeva (2013) for a defence of paraphyletic groups.

caution. Very strong statements about any hypothesis at this evolutionary depth and this stage of investigation are simply going to be wrong, he says (Embley 2012), whether they propose the eocyte tree or the classic three-domain tree. Increasing support for the former gives good reason to doubt the latter (Williams et al. 2013), but nothing is settled yet.

Other fundamental groups?

The three domains, especially when conceived as natural kinds – which evolutionary microbiologists Ford Doolittle and Olga Zhaxybayeva (2013) argue often occurs – are also challenged by genetic exchange between organisms in different domains (see Chapter 3). But these problems do not mean that a default to the prokaryote-eukaryote scheme is warranted either. Especially in evolutionary terms, this division is too crude to capture the process that drove the differentiation in cell types. Eukaryotes are a combination of bacteria and archaea, most noticeably at the genetic level. The cellular structure of eukaryotes, which appeared to be so qualitatively distinct from prokaryote cells to Whittaker, Mayr and many other biologists, is formed from a synthesis of two cell types. Whatever hypothesis about eukaryogenesis (the origin of the eukaryote cell) is eventually victorious (see O'Malley 2010b), the transition from prokaryote to eukaryote has multiple bridging steps that are eclipsed only in the long hindsight of evolutionary events and the inferences needed to reconstruct them. For those who believe a recognizable protoeukaryote came first, with its many accretions of eukaryote features, the transition to full eukaryote status with the acquisition of the mitochondrion is explicable step by step, with many intermediates (Cavalier-Smith 2002; de Duve 2007). For those who see the first eukaryote as the more immediate outcome of the remarkable event of the fusion of two prokaryote cells – an archaeon and a bacterium – (Lane and Martin 2010),[7] the uniqueness of the synthesis still requires a multistage transition as new cellular organization and function evolved after the acquisition of the mitochondrion.

Evolution gives only cold comfort to dichotomies, and even classifications based wholly on cellular organization struggle to maintain the gap between prokaryotes and eukaryotes as a gulf. Many of the 'negative' defining features of prokaryotes (phenomena present in eukaryotes but thought to be absent in prokaryotes) are being rethought in a variety

---

[7] For distinctions between different 'fusion' models, and the debates between their proponents and the proponents of the protoeukaryote hypothesis, see O'Malley (2010b).

of ways. The cytoskeleton, which comprises intracellular structural elements, is now known to be a feature of both prokaryotes and eukaryotes, with many but not all elements of the prokaryote version understood to be ancestral to the eukaryote cytoskeleton (Cabeen and Jacobs-Wagner 2010). Prokaryotes with a membrane around their DNA (the nucleoid) are now well known, especially in the Planctomycetes (Fuerst 2005). Although these compartments are not known to have any shared ancestry with those of eukaryotes (McInerney et al. 2011), they are nevertheless potential softeners of the divide between prokaryotes and eukaryotes. This is especially the case when the division is based merely on supposed absence in prokaryotes of cellular compartmentalization. Various other compartmentalized structures are known in prokaryotes, and these include the magnetosomes of the magnetotactic bacteria discussed in Chapter 1. New imaging techniques have shown the high level of internal organization of prokaryote cells, and how this achieves the isolation of cellular functions. Other classic capacities allegedly separating prokaryotes from eukaryotes, such as the lack of ageing and development into adult forms, have also been challenged by recent findings (see Chapter 4 for a discussion of microbial development). However, energetic constraints persist in prokaryote cells. These limitations prevent major size increases, and might well have been overcome only with the acquisition of the mitochondrion in eukaryotes (Lane and Martin 2010). This acquisition can therefore be seen as a major causal force behind this unique evolutionary transition in cell type. The emergence of eukaryotic cells unquestionably involves the fusion of two cell types, thus returning to the problem of whether monophyly alone can define a natural group (Rivera and Lake 2004).

For some microbiologists, the far more fundamental way in which to carve up nature can be found at the cell membrane level. Early in the official history of microbial classification, distinctions between Gram-negative and Gram-positive cells were made on the basis of staining techniques developed in the late nineteenth century. Cells that take up the primary staining agent, crystal violet, and retain it when an organic solvent is applied, are Gram-positive; those that do not are Gram-negative.[8] Staining properties occur due to cell wall content, and they allow visualization of cells as well as their classification. This division into Gram-positive and Gram-negative is still important in prokaryote taxonomy, but has been

---

[8] The staining process is more complicated than this, with counter-stains and other primary staining agents being used. There are also complications to this scheme due to well-known Gram-variable and Gram-indeterminate organisms (Beveridge 2001).

incorporated into very different evolutionary accounts of fundamental differences in cell membranes. Evolutionary microbiologist Thomas Cavalier-Smith (1987) proposed that the loss of the outer cell membrane in bacteria was the original evolutionary 'trauma' for cellular life. Gram-negative bacteria, with double cell membranes, are thus the pre-trauma ancestors to Gram-positive bacteria and other 'unimembrane' organisms.

Microbial phylogeneticist Radhey Gupta took up the same theme, but used the categories of 'diderm' and 'monoderm' to tell a different evolutionary story. He argued that Archaea and Gram-positive bacteria form a single primary group, which is the monoderms or single-membrane pro-karyotes. The Gram-negative bacteria are the diderms, which have double membranes and arrived later in evolution (Gupta 1998). According to this model, archaea and bacteria are not monophyletic groups, but monoderms and diderms are. This division has molecular and morphological data to support it – not enough of the former for many molecular systematists – and can account for some of the three-domain discrepancies (Gupta 1998). Cavalier-Smith's competing model also emphasizes the defining nature of membranes, and agrees on the importance of putting molecular data within an organismal, mechanistic context (Cavalier-Smith 2010). He suggests that 'bacteria' is the best terminology for all prokaryotes, and should be used in its 'classical' sense to encompass Archaea, rather than falsely separating them from their bacterial evolutionary context.

Making high-level classificatory divisions depends these days on bio-logical differences being explained as the causes or consequences of major evolutionary divergence. Attributions of the degree of importance of those differences – all located amongst microbial life forms – depend at least in part on whether cell biology, biochemistry or genome content are seen as the most fundamental features to track. The point I am making here is that however basic such differences may appear, they have detailed and lengthy mechanistic explanations underpinning them, and that once the detail is understood, the 'fundamentality' of any phenomenon is a more modest outcome within a much bigger evolutionary and explana-tory context. Naturally, the attention such differences attract is very valuable, and the ways in which organisms cluster together over evolu-tionary time a true source of knowledge, but substituting one sort of fundamentality for another in the process of gathering more knowledge seems both historically and epistemically risky. This is historically the case because with increasing information, new hypotheses become possible and these displace and refute the old ones; it is epistemically incautious, because thinking of biology as given to fundamental divides – true natural kinds – is a project littered with failures. The few things that do seem certain are that biological variation exists in a vast array of subtle shades and grades, that emerging and future technologies will reveal more of these subtleties, and that any recognition of important distinctions needs to be tempered with careful argumentation about why and how such distinctions are being made.

The kingdom of protists

It is important to recall as well, however, that classification at the kingdom and domain level is not just about making major divisions between life forms, but also about connecting them ever more closely. This is what can be seen in evolutionary protistology, the science that classifies the unicellular eukaryotes (discussed in the introductory chapter; see also O'Malley et al. 2013). The third domain mentioned above, Eukarya, includes an extensive variety of microorganisms, as well as the various multicellular forms of plants, 'true' fungi and animals. Eukaryotic microbes are predominantly protists, which are the unicellular organisms with non-prokaryote-like cellular organization. They have a nucleus as well as numerous membrane-bound organelles. Protists further complicate general thoughts about what microbes are. They sound like a small add-on group of oddities – and have been for much of the history of taxonomy – but are, in fact, the key to understanding the phylogenetic, biochemical, genomic and trophic (nutritional mode) diversity of eukaryotes.

After two centuries of referring to all unicellular organisms as animalcules or infusoria and placing them within the highest-level groups of plants or animals, a fairly abrupt shift in the middle of the nineteenth century saw several proposals to separate out unicellular organisms. This 'group', which included 'anucleated' organisms ('bacteria') once they were recognized, was called by various proponents Protozoa, Regnum Primigenum, Protista and Primalia (Rothschild 1989). Natural historian Ernst Haeckel (1834–1919) proposed a famous genealogy of organisms in 1866 that contributed to these efforts. He described Protista as consisting of unicellular organisms that while sometimes forming colonies did not form tissues (Whittaker 1969). Herbert Copeland (1902–1968) expanded this kingdom in 1938 within his new four-kingdom system to include the multicellular organisms of algae and fungi. He defined protists as not 'descended from any form which would be properly regarded as either a plant or an animal', but which still formed a natural group (Copeland 1938: 393). Copeland dedicated quite some effort to separating the Monera from the Protista, and establishing the protist phyla. Monera were too sparse to have phyla, he argued; their limited diversity meant they would fit into a few classes.

Considerable objections to the protists as a category of 'intermediate or intergrading' organisms are on record (Moore 1954: 588), and these objections invoke epistemic reasons such as no increase in classificatory power, the obscuration of more important distinctions (plants and animals), and the arbitrariness of putting 'hard to classify' organisms into a bin of their own (e.g., Weller 1955). Whittaker's kingdom categories preserved the protists, as did Margulis's modification. She argued that protists should be explicitly recognized as 'polyphyletic evolutionary "experiments"' (Margulis 1971: 243). Later discussion still sought to establish the group as coherent despite its diversity, and to found it on positive distinctions:

> Protista are neither mini-plants nor mini-animals nor a combination of the two ... [I]ts 120[,]000 species should not ... be viewed as merely a transient stage – or level of organization – bridging the gap between the prokaryotes and the higher eukaryote groups. They have an integrity and taxonomic and phylogenetic cohesiveness of their own (Corliss 1984: 91).

But for most large-organism biologists, such as Mayr, protists were a rather dubious form of life:

> [W]hen you get into the low eukaryotes, there is this group that is a sort of a garbage can called the protists. And there are authors I'm told that recognize 80 phyla of protists. God knows what there is in these 80 phyla. And most of them do not have species in the normal sense. They don't have a proper process of speciation or anything like that (Mayr 2004: unpaginated).

Because many protists are not obligately sexual – although probably all of them are facultatively (see O'Malley et al. 2013) – this means reproductive isolation does not apply in the way that zoologists, such as Mayr, thought it should: as the central mechanism of speciation. Not only does this view refuse microorganisms a place in the modern synthesis of evolutionary biology, but it also misclassifies them in a truly important sense that has major implications for even a zoocentric view of evolution.

The most recent overarching classification of eukaryotic life – one likely to prove stable except for clarification of smaller branches – clusters all eukaryotes into five or six major groups (Adl et al. 2012; Figure 2.5). Plants, animals and fungi are subgroups within the clusters defined by evolutionary relationships amongst the protists. When non-microbiologists reflect on the category of eukaryote they rarely consider its evolutionary and taxonomic dependence on protists. Most eukaryotic diversity is unicellular and the structure of eukaryote groups is determined largely by working out

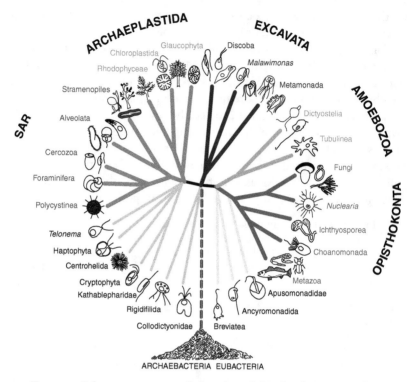

Figure 2.5: Eukaryote supergroups (Adl et al. 2012). Used with permission from John Wiley and Sons.

how protists have evolved their various characteristics. Plants, animals and fungi are simply special cases of unicellular eukaryotes in both classificatory and evolutionary senses (Simpson and Roger 2004; O'Malley et al. 2013). A quick examination of the eukaryote tree helps make this point (Figure 2.5).

The identification of several superkingdoms of eukaryotes is a major phylogenetic achievement that situates eukaryotes such as ourselves within the larger biological scheme of things. All large eukaryotes arose from protistan ancestors. Protists outnumber large eukaryotes, both in diversity and actual numbers. Not only does this tree allow fine-grained classification but it also provides hypotheses about and evidence for the evolution of different eukaryote groups (for example, the nature of the ancestral metazoan; see O'Malley et al. 2013). What this tree does not do is identify the most basal group of eukaryotes: it is an unrooted tree, which means the evolutionary relationships between supergroups are not fully understood. There are, as the previous section noted, ongoing debates about the nature of the earliest eukaryotes, and some systematists – particularly evolutionary protistologists – doubt this debate will ever be fully resolved (see O'Malley et al. 2013).

However, what this classification does achieve, apart from scrupulously organizing groups on an evolutionary basis, is to reveal the diversity of unicellular eukaryotes and the impossibility of classifying larger eukaryotes without considering the protist world. Faced with such a classification, it makes more sense to think of our animal selves and our closest relatives, the fungi, as big protists. Plants too can be understood in this way, and algae are already organisms that are conceptualized as liminal between the kingdoms of protists and 'proper' plants (Delwiche and Timme 2011). The group called the eukaryotes is thus the realm of protists, with some interesting variations on what being a protist is amongst the plants, animals and fungi. Metazoans, for example, can be regarded as 'the simplification of a more universal biology' in protists such as ciliates, 'albeit altered and refined over two billion years of divergence' (Bracht et al. 2013: 413). From a more sociological point of view, many disciplines such as mycology (fungi research) and phycology (algae research) are far more at home in protist meetings and amongst protistology researchers than elsewhere in zoology or botany. This is not to say that individually visible animals, plants and fungi do not matter. They definitely do, but they need to be understood at their most basic from within protist classification and evolution.

Concluding thoughts on kingdoms and domains

These debates about kingdoms and domains could be interpreted not as a rebuttal of fundamental divisions in the biological world, but as evidence against the current categorization schemes on offer. There is no actual 'theory' to underpin why such divisions are natural, apart from standard expectations that any such group should be monophyletic (de Queiroz 1997). In other words, '[e]volution, tradition, and practicality dominate taxonomy at the kingdom level' (Rothschild 1989: 302). Establishing the naturalness of such groups is not simple, as the case of protists shows. Either they are leftovers once the big things are taken out, or they describe all eukaryotes with some special sub-groups constituting the big things. As palaeontologist Michael Benton argues (2000: 643),

> it is entirely unnecessary to demonstrate that higher category names have any fundamental meaning . . . they work, and they offer instant guidance on the hierarchical structure, so why abandon them?

Nevertheless, there is always a craving in taxonomy for more principled justifications for decisions about which organism goes where. Is there any such justification at any level in classification practice that obtains for microbes and other organisms? Many systematists would argue there is indeed, but that this is at the level of species, the lowest rank in today's Linnaean hierarchy and the only one on which evolutionary theory has a direct bearing. This is because other levels are conceived as based on species, and thus gain theoretical justification only indirectly – in standard classification practice anyway. As the next chapter will show, this is problematic, too, because any concept of species cannot reach across all domains of life. Although one solution to this problem has been to exclude the most problematic organisms – microbes – from standard species concepts, the reflections above on kingdoms and domains show how this strategy is very difficult to justify when multicellular organisms can be enfolded within microbial classification schemes.

Microbial classification is not, therefore, just an issue for microbiologists. What this chapter has shown is how the most basic divisions of life – domains and kingdoms – are all dependent on how microbes are classified. Plants, animals and fungi simply have to fit into the relevant groups in the microbial world. The major evolutionary groups are defined microbiologic- ally, and the major events leading to the emergence of those groups (especially the origins of life, the divergence of bacteria and archaea, eukaryogenesis and the many origins of multicellularity) were all microbial events. In other words, the relationship of eukaryotes to the rest of life, and of eukaryotes to one another, can only be systematized on the basis of microbiological knowledge. One response to this conclusion, and a reaction that sometimes philosophers have, is: 'Does *everything* have to be micro- bial?' I tend to reply that there is certainly a lot of biology that does not require an explicitly microbe-centric perspective. That is true of classifica- tion, if we want to look at iguanas and hummingbirds, or sunflowers and truffles, in strict isolation from their broader biological contexts. But if we want to understand more basically and inclusively where these lineage characteristics came from, what larger evolutionary groups these lineages are part of and how these organisms survive ecologically, then we need to take microbial classification into account. And whether we are philosophers or biologists, if we want to think harder about classificatory concepts or frameworks, then there is no choice at all but to consider not only debates in microbial classification, but even the basic theory that underpins today's classificatory practice. The following chapter will emphasize that point in relation to species.

CHAPTER 3

Philosophical debates in species-level microbial classification

The species debate forms a deep undercurrent in microbiology, both historically and in contemporary practice. This chapter will outline those disagreements in light of the microbial characteristics that frustrate universal species concepts and even consistent practical definitions. An environmentally oriented approach to classification is an alternative that emerges in the aftermath of these debates. It points to the causal capacities of larger functional groups of microorganisms.

Species as the unit of microbial evolution

The concept of species has a long and troubled history in microbiology. Microbiologists have made considerable efforts over the last century to find a 'natural' classification system. Scientists, historians and philosophers have frequently remarked about the difficulties of doing so. Without recapitulating too much of this well-known debate, I will summarize the issues and assess where microbiology is today in regard to the elusive concept of species – and more practically important, the operational definitions employed – in order to work out how much this conceptual status matters to classification practice.

A short history of species classification in bacteriology

One way of parsing the entire history of microbiology is as a history of the quest for species and the right species concept. Until the 1770s, says historian Marc Ratcliff, 'animalcules did not exist as *species*, but only as specimens' (2009: 190). Many of the most famous early observers of microorganisms, including van Leeuwenhoek, did not name a single type of microbe. Instead, they referred to the organisms they observed in only the broadest and most colloquial terms. This lack of shared systematic classification is believed to have greatly slowed microscopic investigations in the eighteenth century (Ratcliff 2009). Medic and natural historian John Hill (1717–1775) set out in the early 1750s a template for the Linnaean classification of microorganisms and went so far as to propose 'animalcules' as a new kingdom of life. His ambitious efforts failed to overcome the specimen-based listing of animalcules until 1758, when Linnaeus set out a general category for infusoria (anything unicellular). A few years later, in the 12th edition of *Systema Naturae*, Linnaeus made a distinct category for bacteria, which he named *Chaos infusoria*. The *Chaos* genus contained five species and required classificatory rule bending in order to fit them into the Linnaean system (Ratcliff 2009). It was, therefore, as anomalies and violators of what had already become a well-established classification system that the smallest infusoria (prokaryotes) were incorporated into natural history.

Drawing on and extending Linnaeus's classification of microscopic life from the 1770s onwards, Otto Friedrich Müller (1730–1784) made considerable progress in establishing infusoria as a properly classifiable realm of life (Drews 2000). He categorized eighteen genera of infusoria (prokaryotic and eukaryotic) and set up a framework able to incorporate new observations of microorganisms. With improved microscopes, Christian Gottfried Ehrenberg (1795–1876) elaborated in 1838 a greatly expanded classification system with additional prokaryote genera. He used his classificatory studies as a vehicle to oppose spontaneous generation (the ongoing origination of life from non-living matter), and put enormous effort into the identification in protists of 'organs' analogous to those in animals (Jahn 1998). Further fine-grained scrutiny by microscopists such as Félix Dujardin (1801–1860) produced more encompassing taxa with a greater array of definitive characteristics. These classifications were also critical responses to Ehrenberg's rather dubious (in hindsight) theoretical justifications of his taxonomies. However, for many microbiologists, it is Ferdinand Cohn (1828–1898) who provided the foundation for today's microbial taxonomy. He initially trained as a botanist but became best known for his work on microbial classification, which he had learned from Ehrenberg. Cohn placed bacteria in the plant kingdom – where he included the troublesome fungi – and attempted to find bacterial approximations of genera by providing detailed descriptions to compensate for 'deficits' in microorganismal characters (1875, in Brock 1961).

Cohn's system succeeded by making analogies to plant and animal classification, from which he extrapolated 'form genera' and 'form species'. This method allowed him to suppress any physiological

Figure 3.1: Some major historical figures in bacterial classification: Linnaeus, Müller, Ehrenberg, Cohn, Bergey. Bergey's image is used with permission from the Bergey's Manual Trust.

heterogeneity within those groups. Physiological variants in bacterial groups became 'varieties' or 'races', in the same way they did in plant and animal taxonomy. Genera were broad groups with big differences from other groups, and species were groups or specimens with small, single differences from others in the genus. These differences could then be confirmed as 'natural' by culturing the microorganisms (growing them as pure lines on artificial media – see below) and confirming their characteristics (Cohn 1875, in Brock 1961). Cohn's formalization is the basis for what became the definitive manual of 'bacterial' classification, first implemented by David Bergey (1860–1937) and co-editors in 1923 as *Bergey's Manual of Determinative Bacteriology*, and now continued under the name of *Bergey's Manual of Systematic Bacteriology* (Murray and Holt 2005).

The problem of specificity

The biggest obstacle facing classificatory schemes up to and including Cohn's (and even later, as we will see) was the sheer 'mutability' of microscopic organisms. The key philosophical question asked in relation to classifying microbes was of whether they were organized in any natural sense and could thus be ordered in the same way as other organisms. The possibility of doing so rested on the *specificity* of microbes: whether they had reliable species-demarcating characteristics. Pleomorphy, the existence of multiple forms of the same type of microorganism (with few constraints against taking a new form), represented the competing notion of microbial disorder. It was a violation of the morphological and physiological specificity required to make a classification (Cole and Wright 1916; Mazumdar 1995). Extensive evidence of apparent plasticity, which seemed to allow microorganisms and fungi to change phenotypes repeatedly and drastically over a lifetime, even in constant environmental conditions, was the key reason that microorganisms, especially the 'lower' infusoria (prokaryotes), were believed to be too disordered to be the basis of scientific study and rigorous medical practice.

The fact that for some researchers, 'Bacteria are inconstant and continually lose themselves in one another' (Nägeli 1879, in Farley 1974: 145), meant their systematics needed careful justification:

> One can therefore ask the question if, in bacteria, species occur in the usual sense that we find in higher organisms. Even those who do not agree ... that everything comes from everything and develops into everything will despair when they look into a mass of bacteria. ... However, I have become convinced that the bacteria can be separated into just as good species as other lower animals and plants, and that it is only their extraordinary smallness and the variability of the species which makes it impossible for us with our present day methods to differentiate the various species which are living together in mixed array (Cohn 1875, in Brock 1961: 212–213).

Finding these 'good species' was not an easy task. Early germ theories, such as that of Girolamo Fracastoro (c. 1478-1553) in the sixteenth century, and Linnaeus's musing on microbes as causes of fermentation, fevers and putrefactions (Almquist 1909), were unable to identify specific patterns of causal effects. More developed accounts of microbial causality in the seventeenth and early eighteenth century were unable to establish order in the microbial world. Even in the nineteenth century, despair prevailed:

> The lowest living things are not, properly speaking, organisms at all; for they have no distinctions of parts – no traces of organization . . . not only are their outlines, even when distinguishable, too inspecific for description, but they change from moment to moment, and are never twice alike, either in two individuals or in the same individual. Even the word "type" is applicable in but a loose way; for there is little constancy in their generic characters; according as the surrounding conditions determine, they undergo transformations now of one kind, and now of another (Editorial 1869: 313).

Haeckel complained that 'horrible confusion prevailed' despite huge bodies of data, and was of the opinion that each taxonomist simply made up his own system (1869a: 230). His efforts went into establishing a 'general morphology' that would give Protista and their subgroup, Monera, their own 'natural' historically justified categories. Rather than categorize on the basis of disease-related effects, he attempted to define his groups genealogically, on the basis of 'community of descent' (Haeckel 1869b: 328). But to overcome the changeability of microorganisms, a method capable of firmly subduing their inherent variability and making them constant was required.

The microscopes of the nineteenth century were able to show that protists had diverse but stable enough morphological characteristics, and thus these organisms were more straightforwardly classifiable. Bacteria, however, needed a methodological revolution to be systematized. The investigations of chemist Louis Pasteur (1822–1895) into fermentation convinced him that such processes were the product of specific microorganisms, and that the fermentative capacities of these organisms were stably inherited – part of his argument against spontaneous generation (1861, in Duclaux 1920). Pasteur was making claims about causal specificity with the intent of countering older claims that microorganisms were effects and not causes of fermentation and putrefaction. Even Pasteur, however, was limited by the uncontrolled variability of microorganisms until the pure culture method was developed.

Establishing order in microbial classification

In the early 1880s, Robert Koch (1843–1910), a provincial doctor who was introduced to the scientific world by Cohn, began to follow other bacteriologists and particularly mycologists in the use of solid culture media rather than liquid. By culturing his samples on the surfaces of carefully sterilized potato slices and gelatine[1], different types of organisms could be seen clustered in colonies rather than mixed up in fluid media (Koch 1882, in Farley 1974). Koch's culturing and observation techniques – including staining, slide preparation and photomicrography – were avant-garde. His so-called postulates, formulated to ensure the rigorous association of a particular microbe with a particular disease,[2] demonstrated the consistent effects of different bacterial taxa, despite the sometimes considerable variation or developmental changes within species (Mendelsohn 2002). This uniformity or specificity necessarily had to be passed down from generation to generation, and the pure culture method established that standard inheritance was indeed a property of microbes.

Pure culture thus offered a practical way forward amidst the problems of specificity, predictability and reproducibility of particular microorganismal observations (Gradmann 2000). But by providing a technique that at least partly controlled variability, the tenet of pure culture made bacteria 'monomorphic' (in possession of a single form) even if nature continued to resist this fixity, and 'protean' microorganisms with lifestyle variations and inconstancies continued to be discovered (Cole and Wright 1916). The causal relations that were only hinted at by the revelations of microscopy (for example, Linnaeus's conjectures that animalcula were the causes of disease and putrefaction) could be thoroughly examined in pure culture experimentation that, although simple, was designed to answer a major question about the nature of contagion. Pure culture, with its technique of re-infection from the new culture, was able to establish firm causal links between microbes and specific diseases, transforming some major illnesses from inherited afflictions to contagious ones. It generated reproducible and quantifiable results, even as variability could be tolerated and potentially manipulated as a character itself (Mendelsohn 2002). Not only did such contagions now no longer appear to have purely chemical origins, but also, persisting stable cultures could be examined in a standardized and reproducible way.

The technique of pure culture thus enabled microbiology to designate itself as a true science – one that could order the microbial world with rigorous experimental investigations. The fact that microbiology or bacteriology

---

[1] Gelatine was found to be better than potato because of its transparency, but was quickly replaced by agar-agar because of its higher melting point and resistance to liquification by bacterial enzymes.

[2] Different formulations of Koch's four postulates have been employed, and some historians believe Koch was only indirectly involved in their creation. But their basic form is that: 1) the microbe is present in all instances of the disease; 2) that organism is cultured in pure form in the laboratory; 3) the pure culture causes the same disease when introduced into a healthy host; and 4) the same pure culture can be isolated from the newly infected host.

became a recognized academic field at around this time is a reflection of this achievement, despite such investigations occurring primarily in the applied contexts of brewing, agriculture and medicine. But even with this methodological panacea, cultures could not be interpreted rigidly due to large numbers of laboratory-cultivated colonies exhibiting 'phase variations' that were not the imposed typifying characters of the exhibiting species. But a bigger problem persisted, said bacteriologist Philip Hadley (1881–1963) in the 1930s: 'The chief obstacle in the path of developing a logical, adequate and scientifically exact classification of bacteria is that systematic bacteriologists, possessing no tangible conception of what a bacterial species actually is, experience some difficulty in ascertaining just what it is that should be classified' (Hadley 1937: 185–186).

This problem, pure culture advances notwithstanding, has only increased as more data have been accumulated, and particularly molecular data in the second half of the twentieth century. But curiously, in spite of the difficulties in pinpointing bacterial species, microorganisms nevertheless became model systems for the laboratory study of inheritance and variation in the molecular revolution of the 1940s and 50s. Chapter 6 will discuss how this came about. But for all its initial promise to deliver a natural classification of prokaryotes and other microbes, the molecularization of microbiology has generated even deeper problems for how species are to be understood and microorganisms classified.

Contemporary prokaryotic species definitions

Bacteriology, and microbiology in general, have continued to make considerable efforts to identify species. Although there are some similar issues in the systematics of protists, this chapter will focus on the more extreme difficulties in prokaryote classification. There are strict rules on how an organism is named and the classification validated, with no official status granted yet above the level of class. But while molecular data are now included in any prokaryote classification, the organisms still have to be cultured and deposited in type collections. This requisite step is a tremendous reducer of named diversity for reasons that will be explored below and in Chapter 5. As is the case for broader classification efforts including plants and animals, several operationalizable species concepts have been proposed for prokaryotes, and they mimic those used for large eukaryotes (Ereshefsky 2010). The most widely used definitions and theoretical justifications of species can be summed up as phylogenetic, ecological and polyphasic, and I will discuss each of them briefly below.

But first, the context of these species definitions has to be understood. The modern synthesis of evolutionary biology was built on a particular species concept: the Biological Species Concept (BSC), which postulates reproductive isolation as the defining criterion of full speciation (Mayr 1942). Although there were some problems even early on with its application to hybridizing plants and animals (for example, irises, fruit flies), this concept largely prevails over the evolutionary classification of multicellular eukaryotes. For fungi, the BSC is considerably more difficult to apply, but valiant efforts are nonetheless made. For prokaryotes and most protists, as Mayr and others recognized, there were insurmountable problems with the notion of reproductive isolation. The most major was that of asexuality, which completely undercuts the assumptions of the BSC. If only a single species concept could be endorsed, the conclusion was inevitable: 'All so-called asexually reproducing organisms do not have species' (Mayr 2004: unpaginated).

However, asexual lineages are known to form clonal lineages, and for some systematists and population geneticists, clonal lineages in bacteria seemed to have structural analogies with reproductively isolated ones. Increased recognition in the 1980s and 90s that supposedly clonal populations had 'parasexual' mechanisms[3] and recombinational capacities (Maynard Smith et al. 1993) led to new analogies being drawn with the BSC. It was proposed as an operational tool for classification, based on the argument that recombination within populations should consolidate lineages (Dykhuizen and Green 1991).[4] Findings that recombination occurs less frequently between organisms with very different gene content appeared to reinforce such an argument. But genetic exchange between evolutionarily distant lineages violates these expectations, and this has become a major topic in microbial evolution and systematics.

Lateral gene transfer (LGT), which is synonymous with horizontal gene transfer, refers to three widespread processes by which genetic material moves from one organism (alive or dead) to another. These lateral processes, which are in direct contrast to standard vertical processes of genetic transfer (binary fission, mitotic and meiotic division), are phenomena that

---

[3] Many biologists apply 'parasexual' only to eukaryotes, and particularly to organisms such as protists that have meiosis and some of the other machinery of sex. I use it here in a very broad sense as anything involving the exchange of DNA, solely to save making up another term. It is not unusual for microbiologists to do the same.

[4] It is worth noting that this work about clonal and recombinational patterns comes from microbial population genetics, and the concern is with population structure and not species concepts and definitions per se.

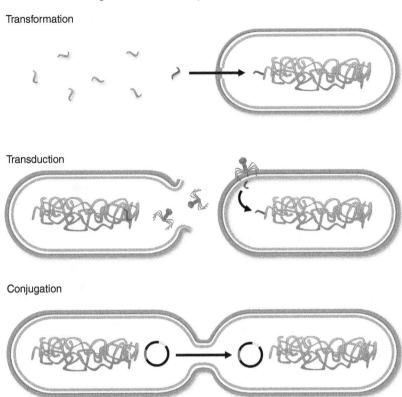

Figure 3.2: The three main mechanisms of LGT. Based on Furuya and Lowy (2006).

have been known for decades and researched intensively in prokaryotes since the 1980s. There is a great diversity of mobile genetic agents and their transfer occurs by at least three mechanisms across small and large phylogenetic distances. *Conjugation* involves cells connecting via a tube-like pilus, which enables DNA from plasmids and other mobile elements to travel from one cell to another (but not both ways); *transduction* occurs when viruses and virus-like mobile elements pick up host DNA during their replication and move it to other cells; *transformation* is the process in which 'competent' cells take up environmental DNA, which is available due to lysis or secretion from another cell, and transport it through the cell membranes and into the cell (Thomas and Nielsen 2005; Figure 3.2).

Following transfer, sequences have to be integrated by homologous, non-homologous (illegitimate), or site-specific recombination processes

(Rocha 2003). Homologous recombination of genetic material (HR) is known as 'legitimate' recombination because sequences have to be very similar at their ends for recombination to happen. It occurs between groups of organisms that are genetically close, but further obfuscates the patterns of vertical (clonal) inheritance presumed to underlie microbial genealogies (Gogarten et al. 2002). This obfuscation occurs not only because LGT means that tracking the history of genes is not tracking organismal history, but also due to the fact LGT can produce patterns that may be interpreted as those of vertical descent (Andam and Gogarten 2011). From a functional point of view, phenotypic properties may be transferred by LGT, especially when different groups of organisms share the same environment. Nevertheless, despite high awareness of the intricacies of LGT and its impact on evolutionary history, the most popular species definition is still a phylogenetic one, and it is necessarily underwritten by the molecules that LGT leads astray.

Phylogenetic classification in microbiology is based on the ribosomal genes identified by Woese as universal phylogenetic markers. They gained this status due to their existence in all cellular organisms, functional indispensability and variably changing regions. These are the genes that encode ribosomal RNA (rRNA); the 16S rRNA gene is the one used for prokaryotes. The 16S gene operates in many respects as a 'barcode gene' for the efficient identification of prokaryotes (Kembel et al. 2012). However, the low resolving power of rRNA sequence is well known: it identifies supposed species by too broad a measure. In addition, classification achieved on the basis of this single molecule says far less than had been hoped about the functional characteristics and other properties of the organisms identified by this marker. And despite Woese's hopes otherwise, ribosomal genes have been caught – both experimentally and via phylogenetic inference – participating in genetic exchanges. All the other genes used to infer phylogenetic groupings are also prone to LGT to a variable and often unpredictable extent, and concatenated averages of genes simply conceal such historical vagaries (Bapteste et al. 2009).

Ecotypes are a more recent way of theorizing prokaryote species and the approach it is part of aims to explain *why* ecologically defined groups exhibit genetic coherence. Methods associated with ecotypes are based on the identification of sequence clusters within communities, in tandem with ecological analyses that attribute 'ecological distinctness' to the sequence (Cohan 2006). The concept itself is intended to capture the maintenance of genomic coherence within ecological populations by processes of periodic selection and genetic drift. This type of selection removes variation *within* the population and increases it *between* different ecological populations, thus forming 'a fundamental discontinuity' between these ecologically separated groups (Ward et al. 2008: 212). The ecotype concept necessarily downplays the extent to which recombination happens in any population, because this would spoil the model (intrapopulational variation would escape purging due to recombination). Increasing findings of recombination within these groups, and lateral transfer amongst them, has proved challenging for the notion of ecotypes as a concept and a tool (Fraser et al. 2007). One interpretation of this contradiction explains the genetic coherence of a population in a habitat as the *product* of recombination (Shapiro et al. 2012). On a practical note, ecotypes are much less inclusive than current species designations, 'which are more like a genus than a species' (Cohan 2002: 457), and accepting the ecotype definition would require a major reconfiguration of current taxonomic practice. Moreover, discerning the 'ecological distinctiveness' of each ecotype is currently done rather coarsely and relies on sequence data to indicate the ecological factors that should be included (Koeppel et al. 2008), thus making the ecotype methodology somewhat circular.

The *polyphasic* approach to prokaryote classification is not really concept-based. It is a pragmatic attempt to integrate molecular and phenotypic data in a way that takes earlier classifications into account. Polyphasic species designations are based on the sensible assumption that single items of information are insufficient for precise and reliable classification, and that diverse data must decide classifications (Stackebrandt et al. 2002). Although initial practice along these lines was not phylogenetic, polyphasic interpretations of genomic and phenotypic data are now always subsumed under a phylogenetic framework. Phenotypic data include whatever morphological features are available (including those exhibited at the colony level), but consist primarily of biochemical, physiological and chemotaxonomic data (the last is about lipid/fatty-acid profiles and cell wall composition).

Pragmatic molecular criteria that contribute to polyphasic classification, such as DNA–DNA hybridization values or ribosomal sequence similarities, are questioned for their adequacy in picking out 'genuine' species as opposed to creating methodological artefacts (Achtman and Wagner 2008). DNA–DNA hybridization is considered valuable for working out where one species ends and another begins, but the precise value (70 per cent reassociation of single strands) is not underpinned by any biological mechanism or evolutionary explanation, and does not relate straightforwardly to 16S similarity measures. On the other hand, although 16S markers are agreed to be insufficient to identify new species, information about ribosomal DNA guides how prokaryote diversity is demarcated in polyphasic practice (Rosselló-Mora and Amann 2001). Whole-genome data are increasingly thought of as superior to these other molecular measures, but they too have serious problems to do with intraspecies variation and how genes have been acquired from other lineages. Polyphasic taxonomy deliberately reflects on classification as a 'consensus' approach (Vandamme et al. 1996), but naturally there are contradictions in data that can only be sorted out by weighting some types of data more than others depending on the organism and classification context. And as some of the approach's major proponents have said, 'There is no way of defining a species that could correspond to both a phenotypically recognized entity and an evolving unit and that will have, even approximately, the same meaning for all bacterial taxa' (Vandamme et al. 1996: 430).

Perhaps the best that can be said about the 'naturalness' of any prokaryotic species designation is this: 'Our credo is that many microbes fall into natural species with cohesive properties, but that one size [of concept] cannot fit all. Species are biological units, some simple and others more complicated' (Achtman and Wagner 2008: 439). Thinking like this helps resolve any tension between the biological world of 'truly' natural species (usually deemed to be animals and plants), and the biological world often thought to have no properly natural evolutionary units – the prokaryotes (e.g., Mayr 1963). However, before accepting this as a solution, it is worth looking at another group of biological entities, but in this case, ones supposedly without 'natural' justifications of what are undeniably evolutionary units. This whole other world of biological phenomena is also missing an appropriate species definition, but in this case, nobody minds if demarcations are made pragmatically.

Virus species

Viruses are the biological entities that use cells to replicate themselves, and which have certain structural features to enable their existence outside and transmission between cells. 'Phage' (from 'bacteriophage') is the specialized term for the viruses that use prokaryote cells. They have a different name because until the 1940s, some virologists and many molecular biologists denied that phage – discovered a quarter of a century earlier – were a type of virus (van Helvoort 1994). Martinus Beijerinck (1851–1931), an early Dutch microbiologist, was the first person to realize in 1898 that viruses were different from other microbes.

He called them 'contagium vivum fluidum' or contagious living fluids. At the same time he reused and made more specific the old use of the word 'virus', which comes from the Latin for poison and initially meant just 'infectious agent' (Creager 2001: 21).[5] Viruses are excellent foci for discussing what the properties and causes of life are, because these entities fall in-between being chemical and fully biological entities. They have a unique capacity to lose all their structural integrity but not their genomic basis, and with their genomes alone as 'instructions' can reassemble a new set of parts manufactured by the host cell (Wolkowicz and Schaechter 2008; see Rohwer and Barott 2013 for a theoretical riff on this capacity). Because viruses do not metabolize or self-replicate, they are for many virologists or microbiologists merely 'borrowed' life – their 'life' depends on truly living cellular organisms (van Regenmortel 2008). Nevertheless, they can be classified, and because of their impact on uncontestedly living things, they are the subjects of intensive biological classification exercises.

There are extraordinary problems involved in classifying viruses. First, there are no universally shared genetic characters (that is, sequence found in all viruses), unlike the ribosomal genes found in all cellular life. In addition, viruses pick up foreign DNA quite frequently, they mutate very rapidly, and are hard to sample effectively and devise good evolutionary models for (Moreira and López-García 2012). Nevertheless, they are classified with the taxonomic ranks used for cellular organisms. A first pass at viral classification can be done on the basis of genome types and is called the 'Baltimore system' after molecular biologist David Baltimore (Baltimore 1971). Virus genomes can consist of DNA or RNA, and those molecules can be single-stranded (ss) or double-stranded (ds). These basic types also have sub-types, such as negative and positive sense strands and replication intermediates. But the main hierarchical viral taxonomy system (www. ictvonline.org/virusTaxInfo.asp) combines genetic information with morphological features about the protein structures (capsids) that encapsulate viruses and phage in their virion forms (the inert viruses outside a cell). These structural features were the basis of the earliest phage classification system, prior to molecular data being included. Further classification criteria for viruses and phage include host range, pathogenicity characteristics, immunological properties and replication strategies.

---

[5] Dmitry Ivanovskii in 1892 had also 'discovered' viruses, but he thought they were bacterial toxins (Creager 2001). Phage were first detected in 1915 and 1917 by Frederick Twort and Félix d'Hérelle.

The main approach to viral classification is not primarily phylogenetic but 'polythetic' or 'polytypic'.[6] This means that members of the group share many characteristics but not all; they form a lineage of genetically related entities, and occupy similar ecological niches (van Regenmortel 2011; www.ictvonline.org/virusTaxInfo.asp). That combination of characteristics requires different methods to identify various aspects of each virus, so classification is similar to the polyphasic classification of prokaryotes – albeit with less phylogenetic interpretation. There are up to seventy properties used as the bases of these polythetic classifications (Ackermann 2011). Just as for prokaryotes, virus species are believed to cluster into higher-level groupings that include genera, families and orders, with a common ancestor at the base of each of these orders. However, it is generally believed there is no 'universal common ancestor' to viruses, with the main groups having independent origins from different cellular lineages (however, see disagreement below).

The unofficial ranks of class and kingdom are not applied to viruses, but in many respects, the distinctions of genome type (RNA/DNA, ss/ds) operate at those levels (Lawrence et al. 2002), and those distinctions are sometimes claimed to indicate fundamental evolutionary divergence in the viral world (Nelson 2004). In another similarity with prokaryote classification, strains or sub-species of virus are recognized, especially when genomic similarity measures are used (van Regenmortel 2011). A shared problem with using genomic data is that interpretations of virus sequence information frequently clash with interpretations made on other grounds (Lawrence et al. 2002). Furthermore, a hierarchical Linnaean scheme is not easily mapped onto the genetic flux and flow of the viral world, particularly for phage. Instead – and this has been argued for prokaryotes as well – a web of relationships rather than a strictly bifurcating tree is more appropriate.

For many biologists, all these factors mean that viral classification falls outside the pale of genuinely biological classification. But if it does, so might prokaryote classification because of the many parallels. The ancient evolutionary past of prokaryotes and even early eukaryotes is opaque to the current phylogenetic methods underlying classification practice. Likewise,

---

[6] The 'quasispecies' concept, which is a calculation made on the basis of the extremely high mutation rates in RNA phage, is not a classificatory concept. Rather than describing phage populations in standard population-genetic terms, quasispecies theory (a mathematical formulation) identifies 'clouds' of mutating sequences and proposes a phage species as a 'complex self-perpetuating population of diverse, related entities that act as a whole' (Eigen 1993: 42). This population dynamic is not, therefore, calculated for taxonomic purposes, and is not relevant to phage or other virus classification (van Regenmortel 2011).

the evolutionary history of viruses is obscure, despite some strong theoretical reasons being postulated for the ancientness of their ancestry (Koonin et al. 2006). The very 'nature' of both prokaryotes and viruses has been famously debated. Microbial geneticist André Lwoff (1902–1994) wrote an essay on 'The concept of virus' (Lwoff 1957) that was the inspiration for the famous 1962 paper, 'The concept of a bacterium', by evolutionary microbiologists Roger Stanier (1916–1982) and Cornelis van Niel (1897–1985). Stanier and van Niel thought the 'natures' of viruses and cells are so different that 'any kind of intermediate organization' is unlikely ever to be found (1962: 19). Lwoff himself argued that viruses were a natural group of biological entities, despite being defined in a largely negative way in relation to cellular organisms: absence of growth, division, most metabolic genes, ribosomes; presence of only one type of nucleic acid; the one positive characteristic Lwoff identified was 'infectious capacity'. The more recent debates about the meaningfulness of 'prokaryote' object to similarly negative definitions, but there are few objections to the general term of virus. This is because it is rarely given any phylogenetic connotations: 'virus' is merely a description of a type of biological entity.

In the microbe-macrobe section in Chapter 2, I mentioned some recent findings of large viruses and the attempts these had sparked to reconceptualize life, or at least conceptualise the virus world more positively and as a parallel form of life. These huge viruses, with over 1000 protein-coding genes, are larger than many small prokaryotes and possess a range of genes not normally found in viruses. These discoveries of vast numbers of new viruses in a wide range of environmental samples have stimulated the revival of older origin-of-life theories (e.g., Koonin et al. 2006) – theories in which viruses used to play a formative role until being seen as cellular escapees rather than precursors to cells. One of these newer accounts sees viruses as coevolving with and evolutionarily responsible for the three-domain divergences (Forterre 2006). However, there is quite some evidence and reasoning against this claim especially about the nature of the ancestral cell that viruses would have had to modify. Moreover, extensive LGT in these giant viral genomes also contraindicates their degeneration from an ancient fully cellular life form (Moreira and Brochier-Armanet 2008). But despite the many parallels and intersections between viral entities and prokaryote life, there is strong resistance to including viruses in universal phylogenies of evolutionary lineages.

There are many good reasons not to include viruses in the tree of life, say evolutionary microbiologists David Moreira and Purificación López-García (2009). The most basic one is that viruses are not alive, and thus contradict the very purpose of the tree of *life*. A less definitional and more substantial problem is that viruses are not a monophyletic group. They have originated at several different times, allegedly from different cell types, and thus have no common origin (Moreira and López-García 2009). However, since all cells very probably have a common origin (although phylogeny cannot locate it), viruses could be mapped onto any reasonable representation of the tree of life (Bamford 2003) and might even help resolve problematic branching points of cellular lineages. The rejoinder to this solution is that viral habits of host jumping and frequent LGT make phylogenetic inferences difficult. Viruses also undergo random genomic changes at rates that are very different from their hosts. These reasons are not entirely dissimilar to those given for excluding prokaryotes from the biological species concept. From a taxonomic point of view, 'It is convenient, and may be justifiable', said Whittaker (1969: 156), not to treat viruses as organisms'. They simply raise too many complications.

Several commentators object to this 'viral exclusion principle' on the grounds of 'false autonomy', arguing that many unicellular organisms and even multicellular ones cannot withstand the tests for autonomy that are imposed on viruses (see Dupré and O'Malley 2009). Others, from a more particular point of view, argue that the giant viruses at least should be included in the tree of life because these forms of life may once have been autonomous and might be very ancient (Claverie and Ogata 2009). Still others argue that proper phylogenetic analyses of viruses are necessary 'to allow virus taxonomy to reflect meaningful biological relationships' (Lawrence et al. 2002: 4892). These issues have not been settled, and it seems unlikely they will be without changes in how people think either about viruses or classification. However, what might be more conclusive is to consider viral phylogeny in relation to uncontested life forms: cells. The history of cellular evolution is represented, via phylogeny, as the tree of life. But classically, this is a tree of species. What happens if species are much more flexible entities than those assumed necessary for the construction of the tree of life?

Lateral gene transfer, the tree of life, and speciation

Methods used to infer LGT have flourished along with more general methodological developments in molecular phylogeny. Gene transfers can be detected through phylogenetic methods that detect similarities and discordances, and through 'compositional' methods that compare sequence patterns (Zaneveld et al. 2008). These methods seek to exclude

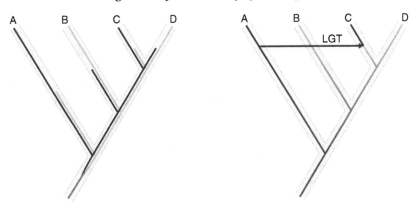

Figure 3.3: LGT (right) versus gene duplication and differential loss (left). While the pattern at the branch tips may be the same, the explanation of that pattern is completely different. Gene duplication and differential loss can be an 'unparsimonious' explanation if many sister taxa would have to lose the gene in different events. Based on Gogarten and Townsend (2005).

other processes that could produce similar patterns to LGT, such as gene duplication and differential loss (Figure 3.3). With less success, these methods try to identify the sources, mechanisms and evolutionary consequences of such transfers, but these are always probabilistic assessments that can incorporate considerable uncertainty (Ragan and Beiko 2009).

LGT in prokaryotes has enormous consequences for the notion of species and the tree of life. Comparisons of strain genomes within a single prokaryote species show that shared genes might be around 80 per cent, but that this number decreases with each subsequent strain genome that is sequenced (Tettelin et al. 2005). Although genome sequencing reveals huge amounts of unexpected diversity, the notion that a 'species genome' exists continues to drive prokaryote phylogeny and classification (e.g., Lan and Reeves 2001: 421), primarily these days in the form of the 'core genome' concept. The idea of a core is meant to deal with the fact that LGT and HR can result in extensive genomic mosaicism within any named species. For instance, a well-known comparative study of *E. coli* proteins in three 'strains' found only 39 per cent of overlap between the three genomes (Welch et al. 2002). And the more *E. coli* genomes that have been examined, the shallower the core genome has become. A 61-strain genome study found that only 20 per cent of the total genes in these organisms were likely to be in every genome. The variable 'pangenome', consisting of all the genes that are not 'core', makes up the remainder (Lukjancenko et al. 2010). If similar rates of LGT and recombination events occur in other prokaryotes – and there is evidence that many are at least as flexible if not more – then an entirely different picture of prokaryote species and their evolution emerges, one in which such groups may be far more horizontally structured than vertically (Andam and Gogarten 2011). Prokaryote species clusters are seldom, as we have already seen above, 'ideal entities with sharp and unambiguous boundaries; instead, they come in multiple forms and their fringes, especially in recombinogenic bacteria, may be fuzzy and indistinct' (Hanage et al. 2005: unpaginated). Moreover, due to LGT and HR, if traditional conceptions of species are used, some organisms may belong to several species simultaneously (Doolittle and Papke 2006).

But because species are the unit of evolution, and the tree of life is a tree of species (or speciation patterns), the problem goes far beyond one of a single isolated concept. The tree of life is the iconic representation of evolution in action; Darwin dreamed of it, and it was supposedly actualized in the molecular era (Woese 1996). The problem with LGT is that individual molecular trees, whether based on single genes or particular combinations of genes, instead of being mutually supportive and providing cumulative support for a unique branching pattern of evolutionary history, frequently contradict one another's evolutionary story. Consequently, many phylogeneticists believe phylogeny needs multipattern webs of life rather than a single-pattern tree of life to represent the plurality of evolutionary histories recorded in the genomes of prokaryotes and other microbes (e.g., Doolittle 1999; Dagan and Martin 2010) – possibly including viruses. Although there may be an underlying tree of bifurcating cells, molecular phylogeny is the only approach with enough discrimination for microbial phylogeny because of the lack of other suitable characters; it is also the dominant approach for large-organism phylogeny. But despite the plethora of molecular data now available, it does not allow the reconstruction of a singular history. And even if such a tree could be reconstructed, it would tell a very limited evolutionary story because it would mask the complex diversity of processes that underlie evolutionary history.

Although there is no doubt that the LGTs that persist in an organismal lineage are infrequent and sometimes invisible, the conceptual implications are inescapable. For universal phylogeny (the tree of the evolution of all life), no gene in any organism can be represented as *never* transferred in the history of cellular life. There is certainly a propensity for some genes to transfer less than others, some lineages to be more recalcitrant to LGT, and some lifestyles to discourage it. But there is no guarantee at any point in evolutionary history that particular stretches of DNA are absolutely resistant to transfer, and that some organisms are completely protected from foreign DNA uptake. There are definitely probabilities and many statistical methods work on calculating such probabilities, thereby coming up with sets of genes that are 'nearly universal' and 'almost never' transferred (Puigbò et al. 2009). But the question here, for both scientists and philosophers, is whether this is the same as discovering the genuine tree of life, or whether it means that such a tree is 'merely' a statistical representation of important trends in the history of evolution. Focusing only on parts of the tree of life, such as the animal branches, or drastically reducing the 'non-phylogenetic signal' of molecular histories, leaves out much of life's evolutionary history. In the end many phylogeneticists trying to get an overall picture of evolution admit the tree of life is a 'pragmatic approximation' and a selective and always contestable pruning of more biologically and evolutionarily realistic networks (Philippe et al. 2011: Box 1).

One of the pruning activities in phylogeny is to focus on so-called core genes – the genes with no lateral history and a universal distribution (Ciccarelli et al. 2006). Unfortunately, such cores are minuscule across prokaryote diversity and produce potentially artefactual trees (Dagan and Martin 2006; Bapteste et al. 2009). Nevertheless, since prokaryote classification is primarily about identifying taxa within a genus, common genes are many more in number than those that would be found across a broader swathe of phylogenetic diversity (Gevers et al. 2005). But there is no such thing as a genus concept or definition either,[7] and there is a degree of arbitrariness alongside hard-earned experience in deciding whether a group is a genus with several species, or a species with several strains (Oren 2004). Perhaps the most valuable insight that could be gained from these problems with the tree of life is not the fact that LGT exists and species might be many different things, but that speciation itself has been too simplistically understood.

Speciation is very difficult to infer in prokaryotes, because if only parts of a genome stop recombining with genomes from similar organisms (see the discussion of HR above), but other parts continue to recombine, then divergence becomes a very messy process (Lawrence and Retchless 2010). Put another way, different genome regions will show different patterns of divergence, and what a tree of bifurcating lineages is representing is lineage divergence that occurs once and for all. And worse, processes of selection, gene flow and recombination work to different degrees in different organisms and environments, meaning that a singular account of speciation will

---

[7] Mayr's effort at a concept of genus states: 'A genus is a systematic category including one species or a group of species of *presumably common phylogenetic origin*, separated by a *decided gap* from other similar groups' (Mayr 1953: 396 – emphases added). This is hardly a rigorous definition, but it does capture standard practice.

fail to capture these often-contradictory processes (Fraser et al. 2009). For now, far too little is yet known about how these processes work together in different lineages with different capacities for speciation.

The way in which modern synthesis architects such as Mayr structured their theory with a particular concept of species has additional implications for how the process of speciation is understood. Mayr and many other evolutionary biologists argued that speciation was primarily allopatric, meaning geographic isolation plays a crucial role (Mayr 1942; 1963; Glossary). But for much of twentieth-century microbiology, microbial populations have been assumed to be wholly *without* geographic boundaries. Almost all microbiologists used to agree with the statement that for microbes, '*Everything is everywhere; but the environment selects*' (Baas Becking 1934; O'Malley 2007 - italics in original). If this were true, it would mean that there were no geographic boundaries to microbial distribution and therefore, no allopatric speciation in microbial evolution.

Now, however, there is growing evidence of the geographic restriction of microbial distributions (prokaryote and eukaryote), although this adds to intra-'species' variation and the overall fuzziness of microbial lineages (Martiny et al. 2006; Chapter 5). But there is also a reasonable amount of evidence for sympatric speciation, in which ecological differentiation drives speciation both in microbes (e.g., Cadillo-Quiroz et al. 2012) and larger organisms. However, both sympatric and allopatric speciation have to be understood in relation to neutral (non-adaptive) processes whereby recombination simply brings about genetic coherence in a population and does not permit divergence regardless of the geographic or ecological situation. These complications are particularly acute in prokaryote populations (Fraser et al. 2007).

The modern synthesis and microbial species

Having outlined these issues – major problems from one angle, but sources of biological and evolutionary insight from others – how are prokaryote species (and for that matter, virus species) related to standard species classification, such as that which goes on in the animal realm? We can turn back to Mayr for his thoughts again on microbial species. Although Mayr knew about the existence of gene transfer mechanisms in unicellular organisms, he never wavered in his belief that bacteria gained their adaptive capacities from mutation (1963: 181). But just as importantly, he thought that the capacities of bacteria for transformation, conjugation and the acquisition of DNA by transduction did not make them sexual reproducers

(1963: 181, 412). Viruses also were poor subjects for the categories of species, he argued, because it seemed – he acknowledged some doubts – they might not really be reproducing sexually (Mayr 1953).[8] 'Only sexually reproducing organisms qualify as species' he argued (1987: 145), and 'some other terminology, for instance paraspecies, will have to be found for uniparentally reproducing forms'. Mayr believed that devising a universal species concept inclusive enough to accommodate asexual organisms would make the concept merely nominal. He found this unacceptable for a key tenet of the modern synthesis. It would be best for evolutionary theory, he concluded, to exclude everything that did not speciate 'properly', even if these organisms were of considerable ecological importance.

Mayr on occasion reflected on the 'observed cohesion' generated by recombination in prokaryotes, but he never approved of attempts to fit the BSC to prokaryotes. He believed that LGT meant microbiology was condemned to a typological definition of microbial taxa (Mayr 2001), and that prokaryotes in particular had to be understood as unnatural groups. He thus followed a fairly common practice of dividing organisms into those that evolved 'properly' and those that could not, and he did this to preserve the BSC and modern synthesis (see O'Malley 2010a). This strategy, however, may have finally reached its limits. There are increasing findings that even in multicellular organisms evolution works on continua of genetic exchange rather than reproductively discrete groups (Mallet 2008), and that LGT affects even large eukaryotes (Keeling and Palmer 2008). In combination with ongoing challenges to plant and animal species concepts, these issues indicate that a gradient of speciation processes needs to be taken into account, even if there are still important differences at the ends of any such gradients (for example, older animal lineages vis-à-vis new ones, or tiny animals such as rotifers that have huge capacities for LGT). Gradients, however, do not tend to produce tidy hierarchies. But this is another way in which the dilemmas of prokaryotic and other microbial classification inform much simpler (that is, zoocentric) ways of understanding evolutionary events and their relationships. One interpretation that will be pursued in subsequent chapters is that by understanding the 'anomalies' in the microbial world, we understand far better the range of processes operating in the biological world in general.

---

[8] Mayr (1953) also encouraged the creation of an international committee on naming viruses (which has come to pass), but thought that there would be far fewer viruses to be named than zoological species.

Environmentally oriented classification: an alternative to the species focus?

But whatever status systematists give to species, it is still very common to ask in biology more broadly how many prokaryote species exist. So far, very few prokaryote species are known and named by the official polyphasic method. By the end of 2013, almost 10,600 prokaryotic species were named and 'validly published' (www.bacterio.net/-number.html). Amongst micro-biologists, however, nobody believes there are this few species, whatever species are. Far beyond the problem of finding the right species definition is that of identifying diversity. Only a small percentage of prokaryotes can be cultured, and if they are not cultured, then that organism cannot be entered into official taxonomic databases. Phenotypic assays (biochemical, chemo-taxonomic) of candidate taxon types have to be performed on organisms in culture. There is now a strong belief in environmentally oriented micro-biology that less than one per cent of prokaryotic diversity has been or even can be cultured (Amann et al. 1995). Cultivation of protists and fungi also occurs well below estimated diversity levels. This recognition of far greater diversity than can be tapped into by conventional microbiological methods, plus the availability of molecular typing methods, has led to an environ-mental revolution in microbiology. I will examine this in greater depth in Chapter 5, but here I want to bring out some of the implications for classification.

Rather than culturing organisms and classifying them solely on the basis of these cultures, metagenomics – the methods with which environmental DNA is sampled and analysed – focuses on microorganisms in their environments. There are different approaches taken, with some focusing on particular genes in those samples (e.g., 16S) and others working with whole genome data (see Chapter 5 for details). Environmental sequences have shown very clearly that even in microorganisms thought to be well understood from laboratory cultures – E. coli for example – there are many cryptic strains with important genetic and ecological differences. These differences are observed even in organisms appearing to be pheno-typically identical to one another when standard species identification criteria are used. Major discoveries of new organisms with unexpected capabilities have been made with metagenomic methods. Most import-antly of all, metagenomics enables microbiologists to examine functional interactions between microbial groups and their environments, rather than having to extract single lineages from those causally complex contexts.

Not only is metagenomics a challenge to the method of pure culture, but even its causal attributions are different, because diseases can be linked to more than a single species. It is now becoming clear to clinical microbiologists that mixed-species communities of microorganisms underpin many diseases (Day et al. 2011; Chapter 5). Culturing is thus doubly displaced from the heart of microbiological classification: first, by molecular sequences and methods themselves, and second, by community-based challenges to its very logic. Nevertheless, culturing remains crucial to the official validation of prokaryote species and other microbial taxa.

Metagenomic classification

Some systematists believe that metagenomic data, because it incorporates genetic and ecological information, can actually give better insight into what species really are and how to define them (e.g., Caro-Quintero and Konstantinidis 2011). Others suggest that metagenomics is not, in fact, about traditional species, nor even about which organisms to put in what classification basket. Instead, it is about functions in environmental communities, and which collectively 'owned' genes and pathways contribute to the maintenance and proliferation of those communities (e.g., Doolittle and Zhaxybayeva 2010). Concerning classificatory concepts, therefore, metagenomics challenges the standard belief that 'species' is the fundamental unit of biological classification and evolutionary continuity. Although isolated lineages will always be important, there is now at least a suspicion that there are multilineal interactive entities with just as much biological and evolutionary 'fundamentality' and causal power. One of them is the community.

The general concept of community is central to environmentally focused classification and research in microbiology. I will discuss in Chapter 5 some of the historical debates that have been associated with notions of community, but here I want simply to suggest that multitaxon groups competing and cooperating within and between themselves are of major importance to an ecological understanding of microbial classification (and even biology more generally). One example of this can be found in human microbiome research. These microbiomes are the DNA sequences of the human body's commensal organisms, which are often called the microbiota. At the molecular level, there is an extraordinary amount of diversity in these communities, especially over time (Caporaso et al. 2011). But microbiota differences within individual humans are less than the differences between the microbiota of other humans, meaning that the community in each individual is highly individualized and even unique (Dethlefsen and Relman 2011). Community structures are often not only distinct and perpetuated over time, but also strongly correlated with particular ecological strategies. These community patterns, which may include host effects, can be modelled above the species level by analysing the 'supra-organismal' functions of the community – particularly the metabolic functions (see Chapter 5).

Microbial communities may in fact drive or at least assist the speciation of their eukaryote collaborators, by creating barriers to reproduction between host organisms and enabling ecological differentiation (Brucker and Bordenstein 2012). Although there is a variety of evidence at both the phenotypic and genetic level for symbionts playing causal roles in host speciation – from endosymbiont genes transferring to the host genome, to hosts losing biosynthetic capacities because they are supplied by symbionts – there is even more evidence indicating that these multitaxon entities can be conceived as units of evolution. These entities persist over lengthy periods of evolutionary time (hundreds of millions of years in some cases) and coevolve with differing degrees of cohesiveness. I will say more about these relationships and units in the next chapters on microbial evolution and ecology. For now, I will focus on the environmental classification done with phylotypes. Phylotypes are an operational unit determined by similarity thresholds of a single molecular marker, usually the 16S gene. Despite all the problems noted above with rRNA genes, they are still used as a proxy for species, or at least to indicate units of ecological diversity (see Chapter 5). In addition, phylotyping and the very basic identification of the components of each microbiome have drawn attention to a higher taxonomic level, which is possibly one of genuine biological organization.

There has been a proliferation of phyla in prokaryote classification, especially in Archaea (Gribaldo and Brochier-Armanet 2012), mostly on the basis of environmental sequence. There are not only no strict phylum concepts, but also no clear rules for their demarcation in prokaryotes or elsewhere. It might be thought that there is no particular reason to focus on phyla for any evolutionary or philosophical reason, apart from this being a handy way to locate broad evolutionarily defined origins. However, metagenomics has revealed some very interesting phylum-level patterns that indicate there may be causal powers to be found at this level.

The implications of thinking beyond species classifications come across very strongly when phyla and core microbiomes are used to predict human health states. Metagenomic studies of obese and lean human twins show that obesity is correlated with phylum-level shifts in the composition of the microbiota in the human gut, particularly in relation to the phyla Bacteroidetes and Firmicutes (Turnbaugh et al. 2009). Moreover, experimental studies of physical transfers of stool samples (and thus microbiota) from obese mice to normal-weight, germ-free mice strongly indicate that these transfers have causal effects. 'Obese' microbiota – the organisms transferred from obese hosts – result in obese hosts, even when diet is held constant. These effects probably flow both ways, with obese human states and the diets associated with them demonstrated to have an influence on microbiota composition – again, at the phylum level.

The concept of a 'core' microbiome that emerges from such discussions is an intriguing one. It does not attribute 'coreness' to particular organisms or individuated lineages. Instead, a set of functions is identified as central to making a living in a particular ecological situation. The core microbiome supplies these functions, which may be fulfilled by very different lineages of microrganisms and shared across a huge range of species (Turnbaugh et al. 2006). Although in any particular community in a specific niche (for example, the human gut) there may be fluctuations over time, there are also mechanisms that restore any diminished core functions after, for example, change in nutritional circumstances or antibiotic disturbance (Dethlefsen and Relman 2011).

These phylum-level effects could be argued to come about because of particular dominant species in that phylum. However, bacterial diversity within each phylum is not what is changing in the experiments. What matters are the relative proportions of phyla to one another (Ley et al. 2006). Although some researchers argue that what should be searched for are specific lineages causing obesity, the core microbiome concept indicates this is not necessarily the right place in which to look for disease mechanisms. Functional effects, if they are properties of communities, will not be located exclusively in individual components of that community. One reason for this might be that LGTs with phylogenetic impact occur below the phylum level. In other words, transfers go on within phyla (or classes), without producing real conflict in the phylogenetic representation at those levels (Andam and Gogarten 2011). These conflicts cannot be absorbed at species-level representations, however, and hence the tree-of-life pattern problems mentioned above. It is in this sense of core functions 'overwriting' species that I am suggesting that metagenomics shifts classification to another focus. Its main purpose is not pinning sequences to trees, or even building trees and hierarchies at all.

But another aspect of metagenomic classification involves an attempt to classify humans on the basis of the microorganisms in their gut. One large-scale metagenomic analysis of human gut samples appeared to show that humans – regardless of geographical location, health status or diet – could be classified into just three 'enterotypes' (Arumugam et al. 2011). These are categories based on intestinal microbiota composition, and were deemed to be non-continuous. The enterotype concept was compared explicitly to that of a blood group, and while not believed to be quite as 'sharply delimited', was suggested nevertheless as able to characterize humans effectively. Three genera dominate these enterotypes: Bacteroides, Prevotella and Ruminococcus. The first two belong to the same phylum (Bacteroidetes) and the third is a Firmicute, but no relationship was found between these phyla and obesity (contra previous studies). However, in a subsequent analysis using long-term dietary monitoring, the first and third enterotypes collapsed into one, and diet was closely correlated with microbiota composition (Jeffery et al. 2012). Considerably more complex patterns of community structure, with continua being far more likely than discrete enterotypes, have since been detected.

There are clearly some major oversimplifications going on in analyses seeking to find discrete human types by detecting genus- or phylum-level microbiota differences, and medical interventions or dietary regimes have not yet been founded on the basis of such classifications. But this practice of coarse-grained classification has been defended on the grounds that classificatory strategies properly able to abstract from complex datasets will be essential for developing metagenomics and its medical benefits. In some respects, however, this medical application of metagenomics perpetuates an old anthropocentric view of microorganisms, in which they are not intrinsically interesting. For example, 'It is what bacteria do rather than what they are, that commands attention, since our interest centers in the host rather than the parasite' (Smith 1904: 818). Metagenomics in general *does* care a great deal about how microorganisms affect us but as much for their sake as ours. In the case of human microbiomes, and even microbiomes in marshes, oceans and soils, researchers might be motivated ultimately by human interests but the answers will still be a lot broader than is achieved by a focus on pathogens and single cultured species.

Ecology and kingdoms

Although these phylum-level based analyses are thought-provoking and perhaps indicative of different taxonomic units that are pragmatically effective for some goals of metagenomic study, there also are implications from environmental perspectives for the three-domain schema. Moselio

Schaechter, a microbiologist who does a great deal to promote the wonders and intricacies of the microbial world via his blog, *Small Things Considered* (schaechter.asmblog.org), argues that as important as phylogeny is, large-scale classifications such as the three domains should incorporate ecological considerations more effectively:

> A sole focus on phylogeny forces the past ahead of the present. Compare the two figures [Figure 3.4]; they represent two different worldviews. One highlights the deep clefts between the three domains; the other is integrative and does away with such barriers (Schaechter 2012a).

Instead of deep evolutionary divides and a hierarchical view of life, Schaechter sees soft differences, extensive continuity and dynamic interaction.

We saw in Chapter 2 some of the challenges to the three-domain tree, and how attributing fundamental differences to bacteria, archaea and eukaryotes is problematic. Although a hierarchical cluster-focused view of life might capture in an abstract and idealized way the evolutionary and similarity relationships between organisms, at best it will capture only coarsely biological capacities for ecological function and organismal interactions. Sometimes, sorting organisms into a taxonomic hierarchy provides a predictive basis for working out what the organisms are doing (for example, whether they are more likely to be heterotrophic than not; whether they will readily exchange DNA or not). But for a closer understanding of what is going on biologically, an ecological and inter-active understanding of microorganisms is needed, not just for microbes but also for all other forms of life. Ecological classification is indeed based on what is happening in the present, as Schaechter (2012a) notes, but it also integrates a large amount of evolutionary information. Patterns of gene transfer, for example, need ecological analyses to explain why – despite genomic diversity – groups of microbes exhibit phenotypic coherence. Past properties and future potential capacities are as much the subject of ecological analysis and classification as present-day approaches, and this combination of timeframes and the capacity for broader explanatory reach is part of the reason metagenomics is growing so rapidly.

The softening of many traditional classificatory boundaries is one of the important conclusions provided by this overview of philosophical issues in microbial classification. Finding discrete cohesive evolutionary groups is achievable with a certain kind of hindsight and idealization, but never via a universal concept or definition. Any such findings are always overwhelmed by the diversity and dynamism of the microbial world, even if microbiology's recent turn to ecological analysis opens up a Pandora's box of issues that hark

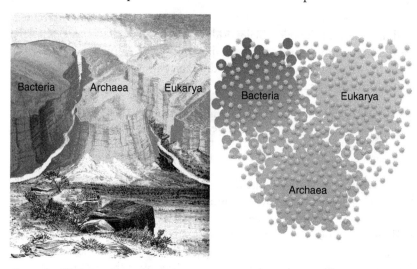

Figure 3.4: Phylogenetic versus ecological representations of the three domains, based on a sketch by Moselio Schaechter (2012a). The many small knobbly shapes overlaying the three domains on the right are viruses. The background canyons are from a public-domain drawing by J. W. Powell (1875, *Popular Science Monthly*).

back to some of the discipline's earliest history. I will explore these ecological developments of the field in more detail in Chapter 5; for now, I will sum up some of the philosophical debates that are ongoing themes in microbial classification.

Concluding thoughts on philosophies of microbial classification

> There is a general feeling amongst taxonomists that if we could find a good general classification – one that is useful and meets the needs of most users – we should be approaching the ideal classification. But while needs change, so do most classifications; what we should not do is make a classification fit a preconceived idea (Cowan 1971: 2).

As Sam Cowan (1905–1976), an eminent microbial systematist noted, natural classification is not usefully contrasted to pragmatic classification. How should we think, then, about 'natural classification'? Does it involve natural kinds? Not unless the very concept of natural kind is massively reinterpreted as it has been, for example, by philosopher Matthew Slater (2014). He defines 'natural kindness' as 'a sort of *status* that things or pluralities of things . . . can have'. A natural kind is thus a means by which understanding can be gained of 'genuine features of the world', because these kinds organize and facilitate the interrogation of that world (Slater 2014: 33). Although this may seem to some philosophers an uneasy amalgam of realism and pluralist constructivism, it does describe well what microbial systematists are doing and believe themselves to be doing, even when they are supposedly at different ends of a spectrum about what 'natural' classification is.

Slater's view of 'natural kindness' helps make great sense of these positions, but it still does not offer much hope for a single unifying classificatory category. Although the most supposedly justified category in biological classification is that of species, there is a great deal of evidence that a universal species concept is neither currently available nor likely to be found in the future. If it ever were, it would be a greatly abstracted one that relies on more concrete and methodologically embedded species definitions, which were themselves designated as needing a single unifying concept because of their incompatibilities. Despite these conceptual problems, taxonomic classification persists, and continues to improve. Virus classification has advanced remarkably over the last two decades. In the protist world, dramatic progress has been made, despite the slower uptake of molecular methods in evolutionary protistology. Even in the prokaryote world, there is far deeper and broader understanding now of prokaryote diversity than there was just a decade ago. Because the straightforward triangulation of molecular data with fine-grained morphological evidence cannot be achieved for prokaryotes, and because they experience a greater incidence of genetic exchange across large evolutionary distances, prokaryotic groups – however they are defined – are much fuzzier than in the eukaryotic world and their designations less stable.

The very idea of a universal species concept is displaced in practice by pragmatic accounts of local contextualized groups. Hierarchies, if they emerge, are not absolute. However, the search for a universal species concept is also a hugely motivating and sharply discriminating tool. It makes scientists check and refine their classifications constantly, not settling easily for 'good enough' classifications. The quest for a universal species concept motivates reanalyses of old designations, accumulations of new data, the development of new tools, and the articulation of alternative theoretical frameworks. The tree of life works as a similar heuristic, illuminating as much through its failure as its successes (O'Malley and Koonin 2011). However, a universally true concept at either end of the traditional hierarchy of life is not indicated by the domains, kingdoms or species stories, and the hierarchy itself can be understood as an imposition on ecological interaction and function: useful for many purposes but far from being the only imposition that has epistemic warrant.

What is ultimately important for anyone doing biological classification is gaining a better understanding of evolved and evolving diversity: what it is, how it functions and how it came to be the way it is. Although it is possible to argue that incorrect species definitions pick up the wrong sort of diversity, in practice this is not what appears to be happening. 'Bad' definitions can be used, results compared and deficient frameworks improved. As Copeland noted quite some decades ago (1938: 384), 'the formulation of a system of classification, then, involves a double set of hypotheses: hypotheses as to the ancestry, origin, and evolution of groups, and hypotheses as to what boundaries will be found expedient'. He discusses convenience very interestingly. There is 'overriding' convenience, which is the necessity of working out which groups are natural; and 'subordinate' convenience, which involves familiarity, conservatism, non-proliferation of higher level taxa and 'feasibility of definition by descrip-tion' (Copeland 1938: 384). This last point refers to an epistemic process by which boundaries are made to coincide with lack of knowledge. What we see in the molecular revolution of microbial systematics are these old limits to knowledge being eroded and with them, what seemed to be clear gaps between life forms. This does not mean, however, that 'naturalness' goes away. All classifications are based on increasingly sophisticated knowledge about the natural world, and making better sense of it is the basic and necessary task of systematics. But as metagenomics has shown, being able to designate taxa and order them on a phylogenetic basis is not sufficient for a broader classificatory view, and a much more dynamic function-based understanding of microbial diversity is now complementing and may even supplant a traditionally hierarchical taxonomic focus.

CHAPTER 4

Philosophical issues in microbial evolution

The microbial world has major implications for how evolution is theorized and researched. Philosophy of biology has focused on the evolutionary theory developed for animals and plants. This chapter will examine what happens to philosophical analyses of evolution when prokaryote evolution is taken into account. I will do this by focusing on several evolutionary processes that are often claimed to distinguish microbial evolution from macrobial evolution. The chapter will conclude with reflections on the ways in which microbial evolution might be the more encompassing way of thinking about and investigating evolution in general.

Microbes and the modern synthesis of evolutionary biology

It is now commonly thought that microbes should fit into the modern synthesis of evolutionary biology, either as simpler versions of multicellular sexually reproducing organisms, or – in the case of prokaryotes – as somewhat anomalous phenomena that do not greatly perturb evolutionary theory. However, this has not always been the case in the history of evolutionary biology. Even now, a number of contemporary microbiologists and other biologists are more inclined to embrace the idea that 'bacteria are different'[1] (Levin and Bergstrom 2000: 6981). They give several reasons for this claim. Prokaryotes and many other unicellular organisms do not need to recombine genomes to reproduce. They commonly multiply themselves through binary fission,[2] thereby giving rise to genetically identical clonal lineages. Because prokaryote genomes are not in a nucleus or specialized germ cell, they are often argued to be 'more directly exposed to environmental influences than the germline DNA of

---

[1] This use of 'bacteria' is an instance of when the word means 'prokaryotes', as I noted in Chapter 2.
[2] There is a variety of modes of reproduction in prokaryotes, including budding, multiple offspring and multiple fission, but binary fission is the standard mode (Angert 2005).

animals' (Sharp and Agrawal 2012: 6142). This exposure makes prokaryote genomes more vulnerable to DNA damage and thus more mutable. Bacteria in particular are known to possess 'mutator' genes, which allow elevated rates of mutation. Prokaryote genomes are also more open to the introgression (incorporation) of foreign DNA.

Microorganisms – again, primarily prokaryotes – have gained evolutionary advantages from mechanisms that allow the acquisition of foreign DNA. These and other microbial capacities are sometimes argued to underpin Lamarckian evolutionary processes in prokaryotes (e.g., Koonin and Wolf 2009). Viruses, even if they are not included amongst the biological entities called microbes, often display remarkable rates of mutation; eukaryotic microbes exhibit a variety of modes for producing genetic and phenotypic variation. Many of these features are conceived of as adaptive, but there is no doubt that prokaryotes and other microbes evolve non-adaptively as well. But whether microbial evolutionary capacities are adaptive or non-adaptive, one question that needs addressing is whether microorganismal evolution is different in any important way from macroorganismal evolution. Prokaryotes are the ultimate testing ground for this thesis, because eukaryotic microbes can be conceived as 'transitional' organisms between micro and macro worlds (Lynch 2007).

Microbes have a long history of being thought of as 'different' with regard to evolution. In Darwin's time, microbes were a problem because of the possibility that in trying to understand the origin of species in microbes, explanations proposing spontaneous generation would not be eradicated in the way many natural historians and experimenters thought they should be (O'Malley 2009). Even after the acceptance in the twentieth century of evolution by natural selection, many biologists continued to believe that microorganisms evolve by Lamarckian mechanisms, and that bacterial forms of life were not appropriate for inclusion in the modern synthesis of evolutionary biology because of their radically different cellular organization and mode of reproduction. Although famous experiments in the mid-1940s were represented as definitive refutations of Lamarckian inheritance in microorganisms, Lamarckian claims continued to arise as biologists tried to get to grips with how microbes evolve and whether standard evolutionary theory could accommodate their obvious differences.

For founders of the modern synthesis, microbes, especially prokaryotes, could be safely and even necessarily ignored. Julian Huxley (1887–1975), in his reconstruction of the modern synthesis, was perfectly clear: 'non-cellular [i.e., viruses] and non-sexual organisms such as bacteria have their own evolutionary rules' (1942: 126) and these were not the rules that preoccupied the architects of the modern synthesis. The 'nature' of bacteria, according to Huxley, is and always would be different, and thus their modes of variation, heredity and evolution. Huxley said this even though he was well aware of the limitations of a narrow organismal focus. He objected to the restricted attention given by evolutionary theorists to plants, for example. Huxley believed this exclusion to be warranted in the case of bacteria, however, because they were thought to have neither genes nor stable heredity (Huxley 1942: 131–2). Somewhat better informed two decades later, Mayr was convinced that the source of prokaryotic genetic diversity was mostly mutational and that LGT was of minor importance (1963). For him, LGTs were irregular events that, like occasional hybridization, evolutionary theory could absorb without difficulty. More problematic, as the previous chapter discussed, was the asexuality that prevented 'proper' speciation and placed microbial evolution outside mainstream evolution. Microbial population genetics, as it developed during the 1970s and 1980s (after an early burst of activity in the 1950s), focused on understanding the special clonal structure of microbial populations.

Despite this tentative consensus about microbes amongst the main advocates of the modern synthesis, there were many intellectual and practical pressures to prevent such exclusion. First, as Darwin had recognized, any general theory of evolution by natural selection had to accommodate microbes or be the worse for it (Darwin 1861; O'Malley 2009). A broad accommodation could be fairly easily achieved (every living thing evolves and natural selection explains important aspects of that evolution), but this sort of inclusion was too abstract to be tremendously satisfying. It became even less satisfactory as molecular evidence – some of it apparently contradicting standard evolutionary explanations – began to pour in from microbes after the 1940s. The third edition in 1974 of Huxley's *Evolution: The Modern Synthesis* made explicit acknowledgement of this new body of molecular genetic knowledge about bacteria (Huxley 2010 [1974]).[3] Much classic research on evolution, ecology and molecular biology had been done by then with microbial model systems, and gaining insight into their evolution and relevance to animal and plant evolution became necessary. As had been recognized even in Darwin's day, microbes are godsends for experimental evolution due to their rapid generation and manipulability (Dallinger 1878; see Chapter 6).

---

[3] This supplementary material was written by microbiologist Mark Richmond as part of the multiauthored introduction to the third edition.

As a consequence of the extent to which experimental evolution has been carried out on microbial and especially prokaryotic populations, questions about evolution can also flow the other way, not only from macrobes to microbes (as was the case for microbial population genetics), but also from microbes to macrobes. But before reaching any conclusions about how microbial evolution influences understandings of macrobial evolution, I will look at how well microbes can be made to fit the standard evolutionary theory that is based on multicellular organisms. To do this, I will address five main themes before drawing them together in a discussion of how a microbiological starting point would affect the general study of evolution.

Directed mutagenesis and 'Lamarckian' evolution

Most evolutionary theory today relies on the existence of genes and presumes their random variation and vertical inheritance. Since molecular biology in the 1940s made clear that all microbes have genes and a process of vertical transmission, it should have been difficult to exclude them from evolutionary theory. However, the way in which mutation works in prokaryotes had been much disputed historically. One of the most renowned experiments in earlier molecular genetics was carried out in 1943 on *E. coli* by molecular biologists Salvador Luria (1912–1991) and Max Delbrück (1906–1981). They contrived an elegant test to show that *E. coli* mutated spontaneously rather than in response to environmental conditions. The latter process was labelled 'adaptive', with the implication of being 'non-random'. Bacterial evolution was thus shown to be non-Lamarckian (Luria and Delbrück 1943). This experiment was followed rapidly by 'crossing' experiments, which were the basic method of standard genetics and previously unobtainable in bacteria, and then the demonstration of the clonal inheritance of spontaneous mutations (Lederberg and Tatum 1946; Lederberg and Lederberg 1952). These findings launched bacteria as model genetic systems, although microbial population genetics needed further technological developments to become fully established in the 1980s (Levin 1981; 2011; Baumberg et al. 1995). But discovering spontaneous mutations far from demolished Lamarckian frameworks for microbial evolution. Molecular genetics had to cope for some decades with occasional flare-ups of this interpretation of evolution.

One of the most heated controversies along these lines arose in the late 1980s when molecular biologist John Cairns detected the reversion of mutations in the *E. coli lac* operon. An operon is a functionally coordinated multigene unit transcribed as a whole. Cairns attributed the reversion to the causal influence of environmental conditions, specifically lactose in the growth media of the *E. coli.* This finding demonstrated 'the inheritance of acquired characteristics', he and colleagues announced: 'Bacteria can choose which mutations they should produce' (Cairns et al. 1988: 145; see below for clarifications of Lamarckian processes). Cairns also saw the ability of bacteria to 'profit by experience' as a blow against reductionism. This claim of 'directed' mutation sparked a great deal of subsequent research and rebuttal (e.g., Lenski and Mittler 1993), with many critics attributing the mutational changes to hypermutation, a multi-mechanism process in which DNA replication and repair become systematically prone to error (Glossary). That switch can be induced by stressful environmental conditions such as starvation, and some of these proliferating mutations – not all beneficial by any means – may generate the genetic diversity able to produce phenotypes that can cope with the changed environment (Rosenberg 2001).

Despite the propensity for mutations to be deleterious, numerous taxa of bacteria possess mutator strains, which indicates there may be ongoing selective advantages to their existence (Denamur and Matic 2006). These advantages are probably due to mutator genes hitchhiking on positively selected genes (Sniegowski et al. 2000). The process of hypermutation is now called 'adaptive mutation' in order to avoid the more dramatic claims about evolutionary mode that are associated with 'directed' mutation (Sniegowski and Lenski 1995).[4] Explanations of hypermutation invoke second-order selection, which is selection for variability that might have a bearing on adaptation, to avoid any connotation of being pre-selected for evolutionary advantage (Tenaillon et al. 2001). There is also support for a 'by-product' explanation, in which elevated mutation rates are simply consequences of the DNA repair breakdowns that occur in stressful situations. Second-order selection and by-production of hypermutation can occur together.

Although for many evolutionists, the directed mutation controversy eventually turned into an endorsement of the consistency of bacterial mutation with Darwinian evolution (e.g., Sniegowski and Lenski 1995),

---

[4] In case there is any confusion with terminology, 'directed evolution' is a human-directed experimental technique that involves inducing mutations in cells and then screening mutated cells for specified functions. Cells with something like the desired phenotype are then selected by humans and made to experience further cycles of directed evolution until the phenotypic requirements are met more exactly. The way 'directed' is used in this technique indicates why alarm bells ring for evolutionists when it is applied to evolution by variation and natural selection.

for some microbiologists the dismissal of Lamarckian processes altogether remains in doubt. These researchers are using a more classic interpretation of Lamarckian evolution, not of organisms 'choosing' their futures, but of environmental encounters affecting heritable material. Woese and colleague Nigel Goldenfeld, for example, argue that horizontal gene transfer of genetic material is 'essentially' Lamarckian (Goldenfeld and Woese 2011). Computational biologists and evolutionary theorists Eugene Koonin and Yuri Wolf (2009) distinguish 'quasi-Lamarckian' processes, such as error-prone DNA repair or environmentally induced conjugation, from processes involving random undirected variation. Although classic Lamarckianism demands a kind of 'reverse genome engineering' by the environment, for which there are few molecular mechanisms, Koonin and Wolf argue that nevertheless some processes meet this requirement. A major one is the CRISPR phage protection system in which responses to phage are 'learned' by the host genome and then transmitted to the next host generation (see Glossary). Many biologists agree this can accurately be interpreted as a Lamarckian process. Another candidate is adaptive LGT (see the following section). Both qualify, argue Koonin and Wolf (2009), as mechanisms for genuinely non-random directed evolutionary change in prokaryotes (see also Shapiro 1995). However, whether LGT operates as a mechanism 'for' anything is questionable and will be discussed below.

One problem with the argument that microbial evolution can interestingly and uniquely be Lamarckian is that standard multicellular evolutionary biology has been argued to have numerous Lamarckian or quasi-Lamarckian processes (Jablonka and Lamb 1995). Another Lamarckian revival in the 1980s occurred in regard to vertebrate evolution.[5] A number of philosophers and historians of science have examined the directed/adaptive mutation controversy. Perhaps its most interesting message is less what it says specifically about microbial evolution, but rather its implications for scientific practice and orthodoxy. A scientific debate such as that sparked by Cairns and colleagues' interpretations, say some critics (e.g., Lenski and Mittler 1993: 1222), 'revolves around competing hypotheses that can be tested by careful experiments'. But many philosophers and historians of science might suggest that other factors, including social and psychological ones, create at least the conditions for such debates and their perpetuation.

---

[5] Immunologist Edward Steele (1981) proposed a Lamarckian inheritance mechanism whereby changes in somatic cells could be passed on to germ cells with the help of retroviruses. Although this was initially a sketchy theoretical account, Steele managed to generate some experimental data that supported the theory, as well as acquire some influential supporters. However, the experimental results could not be reproduced and Steele's conjecture fell by the scientific wayside (Lewin 1981).

It would be a scientific coup to find clear evidence for fully-fledged Lamarckian mutations in bacteria or any other organisms. Moreover, the perpetuation of debates can have very beneficial outcomes even if the motivating claims are misguided – a comment sometimes made in regard to Cairns and colleagues' initial publication about directed evolution (e.g., Roth et al. 2006). It is not at all a sure thing, however, for a wrong model to have beneficial effects, as many historical and even contemporary claims about evolution demonstrate.

Lateral movements of genetic and cellular resources

Although common or garden-variety point mutations and their vertical inheritance are major sources of variation and heritability in microorganisms,[6] microbial evolution is not based on a straightforward history of vertical inheritance, as Chapter 3 showed. LGT and organelle creation by endosymbiosis – the latter the engulfment of one organism by another, followed by the coupling of their evolutionary fortunes – mean that evolutionary accounts of mechanisms of variation and inheritance need to be expanded. Even though LGT is infrequent and organelle-producing endosymbioses very rare, they have major implications for understanding evolutionary process and pattern. In addition to raising problems for how evolutionary processes such as speciation are understood, DNA transfer and recombination raise major doubts about a unique underlying tree of species – the traditional goal of phylogeny as well as its underlying assumption (see Chapter 3). But issues concerning the tree of life and speciation are secondary to this chapter. They are the outcomes or patterns produced by the evolutionary processes that the modern synthesis set itself the task of theorizing. Lateral movements of inheritable resources require amendments to how the evolutionary process has been understood, and these modifications are my focus here.

One of the main questions raised by both biologists and philosophers when LGT is discussed is exactly how much LGT there is. For many commentators, LGT is considered only a small part of what is going on evolutionarily, and can therefore be incorporated easily into standard accounts and representations of evolution. Estimates vary for prokaryote-to-prokaryote rates of transfer, which depend on proximity, lifestyle

---

[6] The genetic diversity of some prokaryotes is generated much more frequently by recombination than point mutation; vice-versa in others (Spratt et al. 2001).

and the extent of barriers to genome integration (Zaneveld et al. 2008). In particular, free-living organisms are likely to have much higher rates of gene transfer than endosymbionts, primarily because in the latter there is a strong tendency toward genome reduction and very low tolerance of DNA acquisition (Kloesges et al. 2011). On the other hand, genes associated with the maintenance of pathogenic lifestyles are the most likely to be transferred (Nakamura et al. 2004). The exact amount of LGT detected depends on which organisms, how far back their molecular history can be traced, and numerous serendipitous events. Stressful environments, for example, can induce enhanced rates of transfer and recombination, as can the mere proximity of potential donors and recipients. But despite the vast amounts of new data on mobile elements and the many mechanisms by which such elements can be mobilized, no precise and phylogenetically wide-ranging quantification has yet been produced.

A variety of hypotheses exist about which prokaryote genes are the most and least likely to be transferred. The so-called 'complexity hypothesis' argues that informational genes (for DNA replication, transcription and translation) are less common objects of transfer than operational genes, which are those that have housekeeping tasks such as maintaining cellular processes (Jain et al. 1999). Informational genes are highly interactive, and thus less moveable and more difficult to integrate than less interactive operational genes. Other research shows that while such properties capture part of the story, the category of 'operational' is too broad. Only some sorts of operational genes are more prone to transfer (Nakamura et al. 2004). More recent studies point to the protein–protein connectivity or connection 'friendliness' of the protein products of genes as the main discriminatory factor in successful LGT. This distinction does not map neatly onto the informational-operational scheme (Cohen et al. 2011). However, in a broad-scale evolutionary and phylogenetic perspective, actual quantities (for example, of LGT per generation) and types of LGT might be irrelevant. What matters from this viewpoint is whether there are in fact genes with *purely* vertical inheritance patterns, going all the way back to the last universal common ancestor. It is very unlikely that there are any, even if LGT events per generation, or thousands of generations, are rare. The extent to which genetic material is shared laterally matters immensely for phylogeny, as the previous chapter discussed.

Whatever the amount of LGT occurring in prokaryotes, however, it is a great deal more than in eukaryotes. Nevertheless, as a consequence of the search for LGT in prokaryotes, increasing amounts of foreign DNA have been found in protists and fungi (Andersson 2009; Richards 2011), and even plants and animals (Keeling and Palmer 2007). There are numerous inferences of transfer from bacterial to animal lineages. Almost all of it occurs when animals and bacteria are in close association, such as in endosymbiotic arrangements (Hotopp 2011). Prominent amongst such relationships are those involving arthropods and the bacterium *Wolbachia*, which is hosted by diverse insects. Large amounts of the *Wolbachia* chromosome have been transferred to a particular insect host (Nikoh et al. 2008). An extended range of eukaryotes has integrated viral DNA into their genomes. 'Endogenized' viral DNA in eukaryotes is not just from retroviruses (Glossary), which can only replicate once they have made themselves at home in a host genome, but also from most other types of viruses (Feschotte and Gilbert 2012). And transposable elements ('jumping genes'), which are bits of DNA that can change position in genomes, are known to be frequently transferred between different animal species (Walsh et al. 2013). The surge in such findings is due to new technologies and closer analyses – many developed first for microbes – as well as to wider acceptance of the evolutionary ubiquity of such processes, and consequently, more deliberate searches.

Although transfer between eukaryotes and prokaryotes is directionally biased, with most going from the latter to the former, there are some intriguing examples of it occurring the other way or even bidirectionally in the same pairs of organisms (Hotopp 2011). At least two bacterial lineages have acquired eukaryote genes for actin and tubulin, which are eukaryote proteins for protein motor and cytoskeletal elements. These genes are expressed (Guljamow et al. 2007). Moreover, not all LGT involves prokaryotes. Some multicellular organisms exchange genes with other multicellular organisms, possibly via viruses or as yet unrecognized mechanisms. Findings of plants exchanging with plants is contested in some cases, but not when it happens between plant parasites and hosts (Xi et al. 2012). In addition, plant-to-fungi, fungi-to-plant, and fungi-to-fungi transfers are all known to occur even if the mechanisms are not yet understood well. Rotifers, which are microscopic but still multicellular animals, have picked up numerous genes for whole metabolic pathways from bacteria, fungi and plants, and large numbers of those genes are expressed (Boschetti et al. 2012).

Perhaps the correct question for an account of evolutionary process rather than pattern is not, therefore, how much LGT there is, but how much LGT is functional. Again, there is increasing evidence of such acquisitions being expressed, functional and adaptive in the acquiring organisms. There is a large number of illustrations of the extent and diversity of these adaptive acquisitions. For example, halophilic (salt-loving) archaea have transmitted genes for salt tolerance and photobiology (the use of light for physiological functions, including sensing and energy generation) to the halophilic bacterium, *Salinibacter rubrum* (Mongodin et al. 2005). The haloarchaea themselves emerged as a group from an anaerobic methane-producing ancestor that engaged heavily in LGT. These transfers enabled the haloarchaea to take up their current lifestyle of aerobic heterotrophic facultative phototrophy (Nelson-Sathi et al. 2012). Gene cassettes in prokaryotes are a well-known example of functional transfers. These cassettes often encode genes involved in producing antibiotic resistance and can slot into or out of a host genome very neatly because of the presence of integrons (integrating or excising bits of DNA, with promoters to express the cassettes in the recipient genome). These and other mobile elements encode functional genes not stably part of any individual's genome. They can be conceived as constituting a 'floating' adaptive resource shared both widely and specifically in the prokaryote world (Michael et al. 2004).

Very often, discussions of the functionality of LGT and HR presume these processes are or should be adaptive, and that this is the reason they have evolved (e.g., Vos 2009). The evolutionary persistence of plasmids (circular self-replicating DNA elements) and the costly 'accessory' genes they carry can be explained as 'co-adaptive' for both genetic element and host (Harrison and Brockhurst 2012). The fact that prokaryotes expand their protein families more by LGT than by gene duplication is believed to have an adaptive explanation (e.g., Treangen and Rocha 2011). However, significant amounts of transferred DNA remain in the genome due to being non-deleterious, fixed by drift and passed over by purifying selection (Koonin 2009). High costs may be incurred both immediately and in the long run (Baltrus 2013). Integrated foreign DNA in individual genomes may be selected for but then cause extinction of the population, especially if there is a lot of DNA transfer (Rankin et al. 2010). If whole genes or groups of genes are not routinely transferred, and fragments of DNA the more likely culprits, the latter are likely to be less functional but also harder to detect (Ragan and Beiko 2009). And even when large amounts of DNA are successfully transferred and integrated into a genome, these transfers may be neutral until the host environment imposes a different selective regime (Omer et al. 2010). Undetected, neutral or deleterious transfers do not necessarily make LGT of lesser importance, however, either for phylogeny or for any theory of evolution that accommodates both adaptive and non-adaptive processes.

But once the issue of non-adaptiveness is considered, it opens up the overarching question of how and why LGT itself evolved. The widely accepted 'progenote' theory presumes that LGT was rampant in forms of proto-life, until these replicating and interacting molecules stepped over the Darwinian threshold by forming cells (Woese 1998). This scenario envisages the existence of LGT today as a declining phenomenon and merely a sporadically occurring vestige of an earlier process. However, other perspectives see LGT itself as a highly adaptive and therefore still a selected mechanism of evolvability. One line of research probes the hypothesis that an ancient history of acquiring antibiotic resistance against naturally occurring, microbially produced antibiotics is the evolutionary force behind gene acquisition mechanisms (D'Costa et al. 2011). But for other evolutionists, the 'simplest' and thus most desirable hypothesis is that the LGT occurring via conjugation and transduction can be explained as the invasion of selfish DNA (Zhaxybayeva and Doolittle 2011). Parasites such as plasmids and viruses may accidentally produce benefits for their hosts, and hosts may fine-tune such advantages, but basic selfishness explains the extent and persistence of LGT throughout evolutionary history.

However, transformation (the mechanism allowing competent prokaryotes to pick up DNA from the environment) is another story because the host controls this process from start to finish. Bacterial geneticist Rosemary Redfield, in an analysis of the evolution of transformation, argues that this process was selected not for its DNA-acquiring capacities, but for its nutrient-uptake capacity: DNA is straightforwardly food for its ingestors (Redfield 1993). She also holds that other processes of gene transfer show no evidence of selection for sex (the randomization of alleles in members of the same group) or DNA repair, but are instead the products of infectious activities that can be explained by the selfish DNA perspective above (Redfield 2001). In this scenario, LGT via any mechanism need not be understood as an adaptation for genetic diversification, but at most as an exaptation (Glossary) if indeed it is ever selected for in this subsequent role of enabling diversification.

Whatever its explanation, the very existence of LGT has some major conceptual implications for evolutionary theorizing. Chapter 3 has already addressed the tree of life issues, but beyond these, LGT also invites conjectures of 'superorganisms', due to the fact that multiple lineages can share a communal genetic resource in the manner of public goods (McInerney et al. 2011; Richards and Talbot 2013). Rather than being exclusively 'owned' by individual organisms and their descendants, genes that are traded frequently and far afield can be thought of as community property. And if the old adage that every taxon of microbes can be found everywhere were true (which it is probably not, but many genes may be – see Chapter 5), then this genetic resource is both global and evolutionarily persistent. This is why the much-criticized notion of superorganism is sometimes invoked in relation to LGT. For a number of evolutionists, the possibility of genomic material being a planet-wide community asset means that the global community should also be conceived as an organism (Sonea and Mathieu 2001; see Chapter 5). This is not meant to be an equivalent of the Gaia hypothesis because these reflections come unencumbered with thoughts about homeostasis and top-down regulation. 'Superorganism' in this context usually refers more pragmatically to large groupings of organisms that can be researched in relation to their environments in localized and demarcated forms. The various notions of community in which LGT is implicated will feature in Chapter 5 on microbial ecology; now the task is to see how cellular resources can also be traded in a similar way and where this takes microbes and other organisms evolutionarily.

In the eukaryotic world, even though LGT occurs much more often than once thought possible, endosymbiosis is more disruptive of vertical patterns and monophyly despite its rareness. All eukaryotes possess mitochondria, which are remnants of acquired whole cells, and many eukaryote lineages have acquired plastids – of which the green chloroplast is just one – in a series of endosymbioses (Archibald 2009). Endosymbiont gene transfer (EGT), which occurs after the incorporation of the foreign cell, means that DNA from the endosymbiont is transferred to the host genome and proteins exported back to the endosymbiont (Timmis et al. 2004). EGT results in mosaic genomes as the host cell genome integrates genes from the endosymbiont. As already discussed in Chapter 2, eukaryotes are genetic chimeras that possess large quantities of both archaeal and bacterial genes. The initial fusion event that gave rise to the mitochondrion created this chimera, and ongoing LGT has added to it. In the general process of evolution, other genes and endosymbionts may be acquired, and a range of genes and even endosymbionts may be deleted. This medley of processes results in a complex overlay of gains and losses that is very hard to read from the evolutionary record, both molecularly and cell-biologically. Endosymbiosis thus reinforces the problems LGT raises for standard evolutionary concepts. Patterns in evolution are not strictly bifurcating, especially at major transitional points (such as the evolution of the first eukaryote cell, in relation to its acquisition of the mitochondrion), and evolutionary change cannot be understood simply in terms of vertically acquired allele frequencies.

The endosymbioses that produced mitochondria and plastids structure eukaryotes in highly recognizable ways. Other non-organelle producing endosymbioses may occur between eukaryotes, or prokaryotes and eukaryotes, but very rarely between prokaryotes. Viral occupancy of cells is only sometimes referred to as endosymbiotic, usually in reference to tightly integrated viral-host systems (e.g., Whitfield and Asgari 2003). As well as major evolutionary transitions (see Chapter 1), endosymbioses enable more 'minor' transitions such as the acquisition of metabolic capacities that the host could not evolve independently. For example, *Riftia pachyptila*, gutless tubeworms in hydrothermal vents, are able to oxidize sulphur, cope with high temperatures and dispose of waste thanks to their endosymbionts (Dubilier et al. 2008). The partnership is obligate for the worm but the symbionts seem to be acquired only horizontally. Diverse organisms in similar symbiotic partnerships dominate the vent habitat. Their endosymbionts enable the multicellular organisms to participate in metabolic activities that are not achievable otherwise, and these capacities and partnerships are reproduced from generation to generation (Wernegreen 2012). Elsewhere in aquatic environments, a massive range of phototrophy enabling endosymbiotic liaisons exists. Corals, which are partnered by photosynthesizing dinoflagellates known as zooxanthellae, are well known as collections of different lineages that function together endosymbiotically.

In the same way that LGT is often conceptualized as a community genetic resource, endosymbiosis can be too (MacInerney et al. 2011). Endosymbioses can provide 'accessory' functions for the host organism, sometimes rapidly, and often have adaptive effects (for example, metabolic, reproductive, defensive) that encompass the entire community of microorganisms and the host. These communities include endosymbionts that are obligate and inherited vertically, or are acquired with each generation horizontally, with occasional swaps between modes of inheritance. Aphids and whitefly are useful exemplars of complex host ecologies and inheritance strategies, since these organisms harbour both a primary endosymbiont and up to six secondary endosymbionts, as well as viruses of the endosymbionts. Symbiont acquisition and gene transfer in these systems occur more frequently when the hosts shift to new environments (Henry et al. 2013), thereby suggesting both processes are adaptive responses.

Obviously, there can be (and theoretically *should* be) a fluctuating tension between host and endosymbiont fitness. Even in mutualistic relationships, endosymbionts can have detrimental effects in many environments on the encompassing organism and vice-versa (White 2011). On the other hand, supposedly parasitic endosymbionts, such as *Wolbachia*, can have close beneficial relationships with some hosts while still being guilty of parasitism (Saridaki and Bourtzis 2010). This conflict and control dynamic makes hosts and endosymbionts useful model systems for macroevolutionary processes, particularly transitions to multicellularity. Endosymbiotic systems have resulted in the transitions of groups of unrelated organisms to single unified individuals – as in the case of cells with mitochondria (all eukaryotes) and plastids (a huge diversity of eukaryotes). These mosaic individuals are also excellent functional systems for developing a phylogenetically broad understanding of multilevel selection.

Evolving collaborative systems and evolutionary individuals

Units or levels of selection have been a topic of theoretical fascination for evolutionary biologists as well as a particular focus of philosophers of biology. Discussions have usually been held in the context of evolving, multicellular, sexually reproducing organisms, with some telling microbiological contributions to discussions of group selection. An integral aspect of such debates is the issue of cooperation. Cooperation usually refers to relationships that have benefits for others; they need not but can be altruistic (West et al. 2007b). However, cooperation does not imply an absence of competition, nor that selfishness does not obtain. Rather, cooperation is a competitive relationship that is understood socially rather than individually, in a hierarchy of such social competition (Foster 2011). The evolutionary analysis of cooperation is concerned with how cooperative relationships enhance the fitness of participants in a way that gives them advantages over other cooperating groups or non-cooperating individuals.

Microbial interactions over evolutionary time can generate insight into both the evolution of cooperation in general and of specific cooperative systems. There is a large body of work on populations (that is, single-species groups) of socially cooperative prokaryotes and how they supress cheaters in the production of public goods. Populations of the spore-forming bacterium, *Paenibacillus vortex*, for example, form elaborate whorls ('vortices'), which require considerable collective signalling and gene regulation to produce (see Figure 4.1). Phenomena such as these have led some microbiologists to propose a distributional range of 'social IQ' across different prokaryote taxa (Ben-Jacob et al. 2004; Figure 4.1). *P. vortex* is at the 'brilliant' end of the distribution, at least three standard deviations above the mean social IQ. Although it should not prima facie be surprising to find complex patterns of evolved social behaviour in microorganisms, this sociality was for many evolutionists 'unexpected', since social complexity was not considered commensurate with 'simple' life

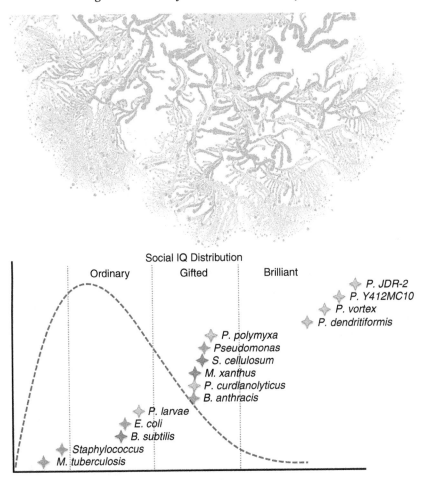

Figure 4.1: *Paenibacillus vortex* and social IQ distribution. Bacterial scores are calculated on the basis of the number of genes associated with 'social' behaviour, which is defined very generally as communication, environmental sensing, and interactions with other organisms including kin. Based on images supplied by and used with permission from Eshel Ben-Jacob, Tel Aviv University.

(Zhang et al. 2009). Models of social evolution were developed in the past mostly for animal interactions, and these are now being transferred with considerable success to microbial social evolution (West et al. 2007a; Velicer and Vos 2009).

Cooperation should, however, be expected amongst organisms that share the same or very similar genomes. It can be explained by kin selection,[7] which for many evolutionary theorists equates formally and empirically to how group selection is understood these days as multilevel selection. The really intriguing question is how organisms with very different genetic endowments work together in ways that are evolutionarily stable and cheater-resistant. I will continue to discuss these relationships as collaborations (see the introductory chapter) in order to capture the range of behaviours that create and preserve such interactions. The mutualistic relationships that exist between multicellular hosts and unicellular endosymbionts have been a major focus of research attention in this regard but there are many other modes of interaction. For instance, swarming groups of the same social bacterium discussed above, *P. vortex*, have been observed transporting the much larger non-motile spores of the filament-forming microfungus, *Aspergillus fumigatus*, from disadvantageous situations to ecologically advantageous ones (Ingham et al. 2011). In turn, the fungal outgrowths (mycelia) create bridges across air gaps that the bacteria cannot otherwise cross. This is despite the fact that these organisms compete for the same nutrients, and clearly requires an explanation beyond that provided by kin selection.

A biologically important model system for non-kin collaborative relationships focuses on the mutualism that has evolved between *Rhizobium* bacteria and leguminous plants. This relationship has altered the phenotype and ecology of the organisms involved (Denison and Kiers 2011). In a history that spans at least 70 million years, plants have allowed rhizobia to invade their roots and obtain carbon in exchange for nitrogen. Fitness benefits are gained by both plants and bacteria, but there can also be costs, especially when cheater bacteria gain access to plant resources and do not give back nitrogen. Some evolutionary microbiologists explain this sort of relationship as mere by-product mutualism that is cost-free and therefore not genuine cooperation (West et al. 2006). But where costs are incurred, additional explanations are required. Mechanisms for repressing defectors are often seen as crucial to the persistence of cooperation and the control of cheaters. The less related the organisms, the more policing is needed. The mutualism between plants and rhizobia is often modelled as a 'partner choice' sanction, in which underperforming partners are cut off in favour of higher benefit ones (Simms and Taylor 2002). But despite the many mechanisms needed to maintain the apparently cooperative multilineage relationships in the biological world, transitions from mutualism to exploitation are surprisingly uncommon (Douglas 2008). This finding is

---

[7] I am distinguishing kin selection and inclusive fitness here, in line with William Hamilton's own expansion of inclusive fitness beyond relatedness (see Gibson 2013: 619).

notable due to theoretical expectations of the selfish reversion from mutualism to parasitism (Sachs and Simms 2008).

Multiple functional relationships, multiple partners and highly variable costs and benefits further complicate explanations of the evolution of mutualistic systems. The relationships between rhizobia and legumes are made more complex by the involvement of fungi, which also gain access to plant roots and often connect them via fungal hyphae to other plants. Large numbers of non-leguminous plants and arbuscular mycorrhizal fungi – fungi filaments that form webs between plant roots and inside plant cells – along with a great variety of endosymbiotic and ectosymbiotic (extracellular) bacteria, form intricate networks of connection and molecular exchange. These networks have persisted since the colonization of land by plants, 475 million years ago (Bonfante and Anca 2009). Indeed, the very existence of these networks may have been made possible only because of the bacteria involved (Naumann et al. 2010).[8]

Guilds, or functionally classified groups, are central to these evolutionary relationships. Rather than tight one-to-one taxonomic associations, guilds exhibit robust patterns of resource use. Despite the taxonomic variability of participating entities, the collaboratively achieved functions in plant-fungi-bacteria systems have been reproduced generation after generation as all partners have proliferated and evolved (Kiers et al. 2011). The situation gains another level of complexity from the fact that each fungal cell can contain hundreds of genetically different nuclei, meaning that at least one of the supposedly 'individual' organisms in the partnership is also some sort of collective (see below). Modes of transmission of symbionts in such systems can be vertical as well as horizontal, but even horizontally acquired partners persist as cooperators in these systems (again, contrary to theoretical predictions). Even though *Rhizobium* bacteria are not transmitted vertically, which is the case in many other endosymbioses, they are encapsulated by host-provided membranes that allow the import and export of proteins, and this shared system is reproduced in each generation of the symbiosis (Bodyl et al. 2012).

Although animals, fungi and plants participate in a great variety of evolutionarily entrenched, multilineal symbiotic relationships, the majority of multispecies symbioses do not include standard multicellular organisms. Multispecies microbial biofilms are a very common example of complex non-kin collaboration, in which particular conditions – especially different

---

[8] In addition, the signalling mechanisms involved in the *Rhizobium*-legume partnership show many signs of having evolved from the much older arbuscular mycorrhizal symbiosis (Ivanov et al. 2012).

metabolic preferences and structured spatial organization – have to apply in order for mutualism to have high fitness values (Mitri et al. 2011). Most evolutionary microbiologists prefer to understand biofilm fitness in relation to the separate clonal lineages comprising the biofilm, with cooperation conceived as occurring within but not between these groups (West et al. 2007a). From this point of view, 'mutualisms are best viewed as reciprocal exploitations' that are not necessarily balanced in terms of benefit (West et al. 2002: 691). However, many instances of multispecies biofilm formation do exhibit potentially costly cooperation between different lineages. Reproductive success for all participants depends on maintaining the mutualism (Leimar and Hammerstein 2010). Even if typical relationships between organisms in different taxa are indeed competitive, an evolutionary account needs to be able to combine competitive and cooperative dynamics to understand the adaptive and co-evolving units involved.

Consider the case of viruses, which can be described as the 'ultimate' example of evolutionary selfishness. They are very successful indeed in the ways in which they gain intracellular access to every form of life in order to reproduce themselves. Many aspects of these liaisons can be explained in terms of selfishness, but numerous other features require additional explanation. A great wealth of data now shows viruses engaged in a range of mutualistic interactions with their hosts (Roossinck 2011). One such case is that of parasitoid wasps, the eggs of which develop within living larvae of other insects. The wasps acquired the capacity to repress the defence mechanisms of their larval hosts from viral mutualists (Renault et al. 2005). The virus replicates only in the larva but is transmitted vertically in the wasp's lineage. The virus and wasp thus form a functional group that is reproduced from generation to generation, with the cost of viral replication and wasp reproduction being incurred by the larval hosts. There are numerous group-forming relationships between bacteria and phage, in which the phage supply genes encoding traits that increase bacterial fitness. These viral relationships may be so intimate that the virus becomes permanently part of the host's survival and reproductive strategies. This is what has happened with the retroviruses that inhabit mammalian genomes. Some of these viruses are responsible for the clade-forming innovation of the placenta (Dupressoir et al. 2012). Viruses therefore might be the paradigm case of how biological entities can function as collaborative agents even when also operating as selfish parasites.

Although it is usually assumed that 'vertical transmission is pivotal for the maintenance of such interactive host-symbiont associations' (Koga et al. 2012: E1230), persisting symbioses such as the *Rhizobium*-legume system raise questions about this assumption. The issue turns on the level or unit of adaptation and whether it is reproducible in any sense. Although many evolutionary theorists may have accepted multiple levels of selection, they still see 'only one level of adaptation – the individual organism' (Wild et al. 2010: E9). I will suggest below and in Chapter 5 that adaptive units and their reproduction can be understood in several looser senses than might be the case when animals are used as models. For other theorists and philosophers of evolution, an appreciation of group dynamics over evolutionary time requires a reconceptualization of what an individual is, and thus what is adapting and reproducing. Evolutionary biologists David Queller and Joan Strassmann discuss the concept of organism in this regard, and define organisms as 'cooperative social groups' (2010: 605) that have controlled or extinguished conflict between group members. These groups can be multispecies groups. Using the axes of cooperation and competition produces four quadrants (see Figure 4.2), each of which can be illustrated with a few examples to show where some of the groups discussed above might fit.

The 'organism' remains the unit of selection, argue Queller and Strassmann (2010: 605), because 'adaptations occur in discrete bundles' that achieve common goals, no matter how many lineages come together to achieve such goals. Clearly, these multilineage organisms can reproduce in a range of modes and yet nevertheless reconstitute the group in a stable enough way to persist in a functional arrangement over evolutionary time. Although 'discreteness' might not quite capture the fuzziness or fluctuating nature of some mutualistic liaisons, this schema nevertheless gives enough flexibility to the concept of organismality in both a functional and an evolutionary sense to make it highly appropriate for many of the microbial partnerships discussed above. Many problem cases can be imagined, but they in fact demonstrate the pragmatic flexibility of thinking of organisms in relation to the dimensions of cooperation and conflict (West and Kiers 2009).

Although it is valuable to set these dimensions out in full (on positive and negative axes of both cooperation and conflict), the collaborative systems that are the topic of this section vary along a continuum that can be cut off the top of Strassmann and Queller's four quadrants (to leave just the organisms-societies spectrum). We could think of the plant-fungi-bacteria associations, which reproduce independently and reconstitute loosely, as one such arrangement. The endosymbioses involving mitochondria and plastids are inextricably intertwined cases of lineage fusion and group adaptation, such that just one individual is presented to selection. At this end – the strongest and most irreversible end of common interest – lie the major transitions in evolution often described as egalitarian (Queller 1997).

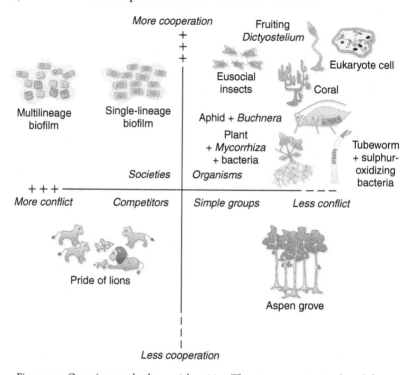

Figure 4.2: Organisms and other social entities. This representation is adapted from Queller and Strassmann (2009), but gives some different placements and examples. It depicts the intersection of cooperation and conflict as axes that demarcate four quadrants of social relationships. Where cooperation is high and competition low, *organisms* are found. It does not matter whether these entities are usually labelled as organisms or not. Where cooperation is still high, but conflict increases, we find *societies*, which are less cohesive and often more fluctuating forms of cooperative social relationship. The other two quadrants of low cooperation and high/low competition are depicted here simply for the sake of completeness but do not feature in discussion in the text. Placements of examples are rough and impressionistic, not carefully calculated (although they may be in Queller and Strassmann's original version).

To describe such transitions of individuality solely in terms of the repression of conflict is inadequate; an account of how common interest evolves is crucial for an understanding of how selection operates on such systems (Leimar and Hammerstein 2010). When common interest is distributed across, for instance, the organisms and viruses in a mutualism so that group fitness sets individual fitness, participants forge an interdependent evolutionary trajectory (however, cf. Gardner and Grafen 2009). This need not involve obligate mutualism – some of the endosymbionts of the mycorrhizal fungi, for example, are able to switch hosts, and rhizobia can make a living outside the plant roots.

Functional integration and collaborative reproduction are key properties of the groups at the high (organismal) end of the spectrum of cooperative systems, and these properties enable these dynamic relationships to maintain functional coherence over evolutionary time. The conundrum of group reproduction in multilineal collectives still needs to be addressed case-by-case and may involve a mixture of transmission strategies (see Chapter 5). However, the very persistence of the many functional groups mentioned above and illustrated in Figure 4.2 indicates that solutions to collective reproduction may be diverse and not understood exclusively via analogies to the reproducing animal. Fitness may also be better understood not via reproductive outcomes but in virtue of the very persistence of these functional relationships (Bouchard 2008). Many such relationships may once have evolved for individually competitive reasons, but the mechanisms of such competition were then 'co-opted' for cooperative functions (Foster 2011). Exploitative and cooperative relationships may persist in these units, both transiently and more persistently, with bursts of rapid coevolution that can refine the symbiosis by increasing adaptive benefits on both sides (Lau and Lennon 2012). This is why it is so crucial to understand mixed-species microbial collaboration: without seeing how it works, we may never understand how cooperation or transitions in individuality evolved, or the units on which selection is operating in particular instances of adaptation. While all of this applies to large-organism communities as well, the microbial instances are not only more common and more pervasive, but also more 'paradigmatic' in a sense I will discuss below.

It is common to think of organisms and units of adaptation in the traditional single-lineage sense because these arrangements are the most easily perceived and cognized, especially in large animals. But today's tools of observation and knowledge making have radically expanded what can be known about the microscopic world.

> Microbiology today is not the discipline microbiologists knew 30 years ago. Everything is new–the technology, the problems that the discipline faces, our training, our conceptual framework, and the way microbiology is perceived by other sciences. Only the organisms remain the same. But is this last even true? (Woese 2006: 14).

Indeed, to answer Woese's question, new metaphysical models of biological individuality may be needed because of the growing understanding of multilevel interactions between microbes and other organisms.

One potentially relevant suggestion is to revive a concept of nested individuals, first suggested by philosopher Gottfried Leibniz (1646-1716). Its central feature is a pluralistic account of biological individuality (Nachtomy et al. 2002). This concept captures in the abstract some of the organismal conjunctions discussed above, and is based on a central claim that the nestedness of levels of individual is a major feature of living entities. However, Leibniz was no evolutionist, and a strict account of nestedness does not fit the fluid examples of individuality outlined above. Another option might be to complement a flexible notion of nested individuality with a framework such as the 'hologenome theory of evolution'. This loose conceptualization is based on the recognition that all animals and plants host microbial symbionts, and these symbionts affect the fitness of the host and the collective. The collective is therefore the paradigmatic unit of selection (Rosenberg et al. 2009). Hologenome theory often sees evolution as both Lamarckian and Darwinian, but not all community-based approaches to evolution include Lamarckian processes.

Godfrey-Smith (2009; 2013) has outlined a multidimensional approach to biological and evolutionary individuality, in which evolution by natural selection operates on different degrees of 'Darwinization' of biological entities. Collectives such as the mutualisms above barely meet his criteria for 'marginal' Darwinian individuals. In many cases, they might be a 'metabolic knotting of reproductive lineages that remain distinct' (Godfrey-Smith 2013: 30). Communities from this perspective might be organisms in a general sense but will not be selected at that level. It does not seem that many of these cases are going to be clear-cut, however, and the wealth of evidence for long-lasting collaborative arrangements indicates that the degree to which adaptive and evolutionary fates are entwined might be diagnosed differently if 'metabolic knotting' is the focus rather than single-lineage reproduction.

A more metabolism-oriented account of individuality can be found in an eco-microbiological point of view, which sees biological individuals as complex associations that involve the collaboration of a diversity of lineage-forming entities. This account will be elaborated in Chapter 5, but the main point to make now is that the object of selection is a single-lineage entity only from a very narrow view of evolution and biological interaction. Selection conceivably operates on units that are formed from the collaboration of many different lineage-forming entities, even in the case of familiar 'individual' organisms such as pea plants, beetles or humans. All of the lineages that constitute the metabolic, reproductive and other physiological functions of the entity have to be considered as part of the adaptive and evolving unit

(Dupré and O'Malley 2009). Biological individuals are probably all situated at the confluence of several lineages of traditional organismal entities, and the adaptive benefits and biological constraints conferred by such arrangements underpin whether these meta-entities or their contributing elements constitute evolutionary individuals. Collaborative ensembles, however and wherever they have evolved, are ubiquitous in evolution, and the capacity to form them is characteristic of all organismal groups and lineage-forming entities. Phenomena such as the egalitarian major transitions in evolution (the ones not about kin) require a better understanding of how such collaborations become fixed (Szathmáry 2012). Amongst the huge variety of collaborative relationships existing now or in the evolutionary past, microbes – especially prokaryotes and viruses – are invariably involved, and the same cannot be said of multicellular organisms, which nevertheless are always themselves part of some collaborative association or other. Microbial systems thus have to be the entry-point into understanding evolutionary and biological individuals, and the relationships between them. This is one of the more basic points of the book and I will flesh it out further in the following chapters.

Microbial development and evo-devo

So far, it could be concluded that microbes are crucial to a complete understanding of evolution. However, there may be major aspects of evolution that cannot be understood from a microbiological perspective. One of these, surely, is the rising science of evo-devo. On one level, it is self-evident that there is no microbial evo-devo: it is ruled out by definition. Only multicellular organisms have differentiated cell types, and development entails this differentiation, so ipso facto evo-devo can only be about those kinds of organisms. Indeed, it has so far mostly been about metazoans, with plants much less of a focus. The lower regard of evo-devo for plants is sometimes attributed to misconceptions about their 'lesser' complexity (Vergara-Silva 2003). A more probable explanation of this skew, however, is because of evo-devo's legacy from animal morphology and embryology. Philosophers have paid a lot of attention to evo-devo and developmental processes in general, and they, too have mostly taken an animal focus, sometimes with awareness of what is being left out.[9] Many of the strongest scientific and philosophical advocates of evo-devo think that the field has challenged the orthodoxy of evolutionary theory, and that major modifications of the modern synthesis are required to accommodate it.

---

[9] See Griesemer (2000) for an effort to conceptualize reproduction and development more broadly.

The claim that evo-devo has radically expanded the modern synthesis – especially in light of the claim that evolution is driven by changes in developmental gene regulation rather than the evolution of new structural genes – is made despite the fact that evo-devo is almost entirely concerned with the evolution of animals, and animal morphology specifically. Evo-devo in this guise thus perpetuates the restricted focus of the older modern synthesis.

As an aside here, it is worth wondering why philosophers like development so much. In part, it is because of the antireductionist tendencies of philosophers of biology. Philosophical antireductionists reject the fundamentality of genetic explanations and search for a viable alternative. Although holism per se has largely been discredited, developmental biology can be regarded as a whole-organism lifecycle approach that produces good scientific results. Even though much contemporary developmental biology is done on the basis of molecular analyses, philosophers of developmental biology see the field's overall research goal of explaining development as transcending these reductionist methodologies. The fact that so much about microbes, including their evolution and physiology, has to be understood primarily on the basis of molecules should be considered in the same light. And if microbes do indeed undergo development, they should be good candidates for a basic account of evo-devo, which would then incorporate animal and other multicellular evo-devo.

There are several ways in which to think about evo-devo in the microbial world and the first is the most obvious. When multilineage systems that include animals, for example, are examined from an evo-devo perspective, that perspective has to account for the development that occurs as part of an evolved process of interaction between the microbial partners and the larger organism (McFall-Ngai et al. 2013). Gut microbes influence the development of the animal gut; arthropod and nematode reproductive systems develop in dependence on bacterial *Wolbachia* endosymbionts; the light organ of the squid, *Euprymna scolopes*, develops dramatically in response to colonization by *Vibrio* bacteria; rhizobia affect the development of plant root nodules even as the bacteria themselves develop into the nitrogen-fixing form of bacteroids in response to signals from the plant roots. Continuing the earlier example, the long evolutionary success story of plants is profoundly illuminated by an evo-devo perspective of the mutualism they have maintained with mycorrhizal fungi and their bacteria (Bonfante and Genre 2008). Evo-devo thus provides another indication that these collectives are indeed units of adaptation and evolution: what develops together has most likely adapted or at least coevolved together.

Understanding such collaborative co-development from an evolutionary point of view is only a modest extension of existing multicellular evo-devo. Although potential objections might be that the bacterial interactions are 'merely' environmental stimuli, case after case of co-development over evolutionary time makes that seem too shallow an answer. Indeed, many multicellular organisms appear primed to 'invite' microbial communities into their developmental cycles. The acceptance of this invitation, often with a chemical password, triggers development in *all* the organisms involved.

But how can development be a capacity of the unicellular partners in these mutualisms? If development is about differentiation of cell types in multicellular organisms, how could it occur in single cells? Stepping outside a zoocentric position, it is easy to find in microbiology a well-recognized phenomenon called microbial development, and a small but flourishing field called microbial developmental biology. These studies define microbial development as 'cellular differentiation in response to changing environments' (Straight and Kolter 2009: 101). It would be considered very different from other organismal development only if the notion of a deterministically unfolding genetic programme were invoked. In earlier decades, microscopic observations of microbial life readily revealed that bacteria both individually and in populations had clearly defined 'developmentally programmed' life cycles, including stationary phases – in which the population does not grow and individual cells undergo major physiological and morphological changes – and a death phase (Finkel 2006).

As with metazoan evo-devo, developmental insight into microbes took off because of molecular evidence. In the early 1980s, geneticists David Botstein and Russell Maurer outlined a number of 'genetic approaches to the analysis of microbial development'. They discussed development as 'complex ordered processes' that were amenable to genetic analysis, regardless of whether they were called development or not (Botstein and Maurer 1982: 62). Although their focus was primarily methodological, they also recognized that there were several instances of developmental processes studied in microbes already: the aggregation and fruiting development of the social bacterium, *Myxococcus*, and the eukaryote slime mould, *Dictyostelium*; transformation into spores in a number of prokaryote and eukaryote microbes; and even the morphogenesis of various viruses. All of these examples are about very different kinds of development, and each of them greatly expands the scope of evo-devo.

One of the most well-known types of microbial development is concerned with the many single-taxon lineages of microorganisms that form multicellular collectives. These collectives develop into entities with differentiated cell types, which undergo regulated coordination and produce different developmental phases that encompass the collective. Early explanation of these 'developmental trends' emphasized that 'the specific bacterial entity is not the cell, nor any colony of cells, but . . . the cell-composite embracing them all in a more or less continuous development' (Hadley 1937: 184). In addition to the well-known model systems of *Myxococcus* and *Dictyostelium*, there is a plethora of other examples of multicellular microbial development. For example, filamentous cyanobacteria differentiate from standard photosynthesizing cells into three other cell types, which include nitrogen-fixing heterocysts, spores and dispersal cells (Flores and Herrero 2010). *Streptomyces*, a soil bacterium, also forms filaments and undergoes highly coordinated development into several cell types with phenotypically important distinctions in physiological, morphological and reproductive functions (McKormick and Flärdh 2012). The gigantic sulphide-oxidizing bacterium, *Thiomargarita*, has two cell types, one of which has a spherical form and a free-living lifestyle, and the other a stalk-like form that clusters with similar cells to form a collective. The different cell types exhibit different reproductive strategies – standard fission for the former and budding for the latter (Bailey et al. 2011). Numerous fungi, including the yeast *Saccharomyces cerevisiae*, can switch developmentally from unicellular to multicellular forms (Brückner and Mösch 2012).

A more controversial example might be multilineage biofilms, which exhibit all these developmental processes of differentiated cell types, communication between and coordination of those cells, and sequential display of new morphological and physiological capacities (O'Toole et al. 2000). Developmental changes are regulated by cooperatively produced signalling molecules in the population or community (Straight and Kolter 2009). And in both single and multilineage biofilms, there is mounting evidence of apoptosis or programmed cell death (Bayles 2007). Apoptosis is a standard feature of developing metazoans and other multicellular organisms, and occurs when individual cells are 'sacrificed' in the formation of new tissues and organism-wide systems such as immunity. Despite all these analogous phenomena, doubts persist as to whether the cascade of genetic pathways for biofilm development is ordered 'deterministically' enough to deserve the label 'development', and whether cell death processes are really regulated and adaptive. Even biofilm development researchers are concerned these collectives might not be experiencing a process controlled by 'a greater genetic programme' that determines 'progression through multiple stages of development' (Monds and O'Toole 2009: 78). Although this seems a rather strong characterization of even metazoan development, the main worry about biofilms exhibiting development is probably that it would involve cooperation. Conceptualizing mixed and single-lineage biofilms as developing entities and units of adaptation requires careful analysis of the competitive and cooperative interactions that produce development. The evolution of group structures and developmental capacities in these collectives is unlikely to be as tightly controlled as that of metazoans.

The developmental biology of such cell collectives stretches the development paradigm quite a bit, but it is equally extended by the concept of development in single unicellular organisms. One model system for this type of development is the bacterium *Bacillus subtilis*, which sporulates via cell differentiation in response to environmental cues. Sporulation results in endospores, which are dormant spores that are highly resistant to environmental damage. Additional cell types can also be produced when *B. subtilis* forms multicellular collectives (Lopez et al. 2009). Another model system for prokaryote development is the aquatic bacterium, *Caulobacter crescentus*. It divides into two cell types at every generation: one a larger sessile (stuck to a surface) stalked daughter cell that can replicate its DNA and divide; the other a smaller swarming (mobile) cell that must develop into a stalked cell in order to reproduce (Kirkpatrick and Viollier 2012). In the protist world, a remarkable illustration of development can be found in *Physarum polycephalum*, an amoeba that forms several cell types. It undergoes a regulated sequence of developmental transformations after a cell with a single nucleus mates or transforms asexually. That cell becomes a huge plasmodium, which is a stretched-out cell with multiple nuclei (similar to the multinucleated fungi mentioned above). In stressful circumstances, the plasmodium can transform into a mass of fruiting bodies in which spores are formed. These can become two types of cells that make further developmental transformations (Bailey 1997). Numerous other protists go through multistage life cycles that are commonly discussed as developmental, and these cycles are both internally and externally driven.

And just to push the envelope a bit further, we can consider viral development, the last of Botstein and Maurer's examples. 'Phage development' flourished as a subfield in an earlier era of molecular biology, right into the late 1970s or early 1980s, as an attempt to understand the regulatory processes underlying life-cycles of various phage. These days, 'development' is used less exuberantly in regard to phage and other viruses, but knowledge of phage 'life' cycles continues to inform the study of the sequential expression of genes in the production of orderly phenotypic change (e.g., Oppenheim et al. 2005). Naturally, this broad use of the term development does not match any definition that mandates development as a transformation from egg to embryo to adult, but this is only a problem for a tightly restricted account of development. The relevant question is whether an animal-exclusive understanding of development is what philosophers want to work with.

This material above makes it uncontestable that there is a large and nuanced body of work on microbial developmental biology. But is there any microbial evo-devo, in which the development of microbes is studied in an evolutionary framework? Most certainly. A favourite evo-devo model in microbiology is again *B. subtilis*, the sporulation of which is regulated by a cascade of genetic interactions. Because of the whole-genome sequences available in microbes, comparative analyses of a phylogenetic range of such regulatory processes can be carried out very effectively. Patterns similar to those of metazoans, such as network conservation, hierarchy and modularity, have been identified in the evolution of the regulation of prokaryote and even virus development (de Hoon et al. 2011; Oppenheim 2005). This is very similar to the sort of work that is going on in today's animal evo-devo, where, for example, developmental gene regulatory networks of sea urchins have been carefully elaborated and then analysed comparatively in relation to the gene regulatory networks of other metazoans (e.g., Davidson 2010). Although the regulatory architecture may be different, and certain 'essential' properties associated only with components of metazoan gene regulatory networks (Carroll 2008), it seems arbitrary and artificial to restrict evo-devo to limited molecular structures and the development of anatomical features (Alonso and Wilkins 2005).

Thus, a major conclusion to draw yet again from these similarities in an evo-devo approach to the macrobial and microbial world is that supplementing mainstream metazoan evo-devo with microbial and other eukaryote evo-devo would enable a far greater scope and depth of evolutionary change to be examined. Microbial evo-devo can also cast light on evolutionary transitions. One example is *Dictyostelium* and its major developmental transformations from solitary single cells to part of a differentiated multicellular body (Schaap 2007). Another is *Volvox carteri*, a eukaryotic microbe that differentiates into immortal reproductive cells and mortal (but independently mobile) somatic cells. Both illuminate different strategies for reaching multicellular individuality (Sterelny 2006b).

However, another aspect of microbial development that might also be relevant to group selection and evolutionary transitions is not concerned quite as much with the evolution of regulatory networks as it is with the evolution of a capacity that allows organisms to 'develop' from one phenotype to another. Phenotype switching is a capacity of genetically identical prokaryotes, and it enables them to avoid the fates of neighbours who do not switch. For example, by stopping growth and replication, and entering a state of antibiotic resistance after signalling between cells, a small number of cells in a population may survive antibiotic environments (Balaban et al. 2004; Kint et al. 2012). As the environment changes, these 'persistor' cells revert to a 'normal' phenotype and their offspring are also unexceptional (Levin and Rozen 2006). Developing from non-competent to competent states (the capacity to take environmental DNA into the cell; see Chapter 3), which is an additional shift *B. subtilis* can make, is another example of this type of development (Johnston and Desplan 2010). Again, this sort of developmental capacity incurs a cost to the group in good conditions but produces a benefit in hard times in that it ensures group survival (Rainey and Kerr 2010).

But switching is not in many cases purely a response to environments: it occurs stochastically, and produces fluctuations in phenotype that may have occasional survival advantages (Balaban 2011). 'Form switching' from a mutualistic to a pathogenic state is one such capacity (Somvanshi et al. 2012). Is this a different kind of development and evo-devo altogether? The ability to switch has evolved and been conserved evolutionarily. It is often interpreted as an evolutionary bet-hedging strategy, in that it creates a small range (sometimes just a binary state) of discrete phenotypes able to respond differently to unpredictable environmental fluctuations (Rainey et al. 2011). Stochastic mechanisms of developmental plasticity are a recognized phenomenon in the development of multicellular organisms (especially insects), where such plasticity is considered to be an adaptive response to unpredictable environments (Johnston and Desplan 2010). Some researchers are now suggesting that stochastically induced pheno-typic change in prokaryotes can be used as a model for the evolution of developmental mechanisms in any organismal lineage, prokaryote or eukaryote (e.g., Eldar et al. 2009). An interpretation that goes even further suggests that developmental switching mechanisms in prokaryotes can be studied for their relevance to group selection because they entail a means of group control and perpetuation (Rainey and Kerr 2010).

Sometimes the transferability of microbial insights to macrobial evo-devo is predicated on the processes involved – involving some sort of unity or continuity of process – but much more often, on the grounds of the tractability of microbial systems for evo-devo study (see Chapter 6). As noted, a great deal of animal evo-devo is nowadays carried out on the basis of transcriptional regulatory networks, and microorganisms also possess such networks; these networks have obviously evolved and often produce adaptive phenotypic modifications. Despite important differences in the evolved structure of transcriptional regulatory networks in bacteria, findings about the gene regulation networks that underlie processes such as sporulation or even pathogenesis are also believed to be transferrable to research on the evolution of similar developmental networks in other domains (de Hoon et al. 2010).

It is probably easy to agree that understanding what is different about the evolution of microorganismal development is crucial to the integration of microbial evo-devo and standard macroorganismal evo-devo. No matter how much or how little difference there is ultimately between the two, comparative knowledge is very unlikely to hinder advances in the traditional form of evo-devo inquiry (see Love and Travisano 2013; Chapter 6). And very interestingly, if group selection in microbes were to be understood as dependent on the evolution of developmental mechanisms to control social behaviour and maintain the existence of the group (Rainey and Kerr 2010), then a broader understanding of development would allow further insight into group selection (see also Griesemer 2000). For example, development in bilatarians ('eumetazoans' or 'higher' animals) might be the product of a conflict-repressing mechanism that arose in the early evolution of these organisms, because development reduces the time in which selfish cells can compete against one another (Frank 2003). Knowledge about even more ancient mechanisms in the unicellular ancestors of these early metazoans will illuminate not only evo-devo, therefore, but also cooperative evolution. The ability to function as co-developing and coevolving groups that straddle lineages and a variety of circumstances means that the microbial world has barely been tapped yet for what it can say to evo-devo and evolution more generally. Just a glimpse of these different mechanisms of development and evolution might indicate that microbes in general, and prokaryotes and viruses in particular, are amongst the most evolvable entities on the planet.

Evolvability

Without exception, the data gathered from in vivo evolution experiments highlight the rarely quantified evolvability of bacteria. These organisms are highly evolvable biological systems with astonishingly fast adaptation capacities (Hindré et al. 2012: 361).

It is possible to think that even if individual evolutionary mechanisms do not definitively distinguish microbial evolution from macrobial, perhaps the combined effects of different mechanisms produce different evolutionary capacities. One of the main candidates for such a difference is evolvability, which is usually understood as the ability of a lineage to evolve by generating more adaptive variation (see below for problems with this definition). Some evo-devo-ists, for example, think that their field is very specially about evolvability, but it has been already pointed out that these claims tend to encompass only a limited amount of evolution (metazoan evolution and development). Is evolvability a good claim to make about any organismal lineage's evolution? Is there any special evolvability in microbes? Prokaryotes and other microbes have capacities to exploit ecological niches and respond rapidly to environmental stress, and these capacities could possibly make these organisms qualitatively more adaptive and evolvable than larger organisms.

As the previous sections in this chapter outlined, microbes have multiple ways in which they can generate genetic diversity that might become adaptive. Hypermutability, gene exchange and endosymbiosis can all be argued to support the evolvability claim. Additional microbial phenomena at the individual organism and population level are also thought to add variability and potentially evolvability. For example, yeast prions (self-propagating protein isoforms) can reveal 'hidden' genetic variation. If these variants are adaptive in new environments, then the organism is likely to survive and proliferate. Prions, therefore, have been argued to provide extra mechanisms of evolvability in microorganisms (True and Lindquist 2000). One problem for any assessment of microbes as more evolvable on the basis of greater genetic variability is that sexual eukaryotes, including unicellular ones, are themselves meant to have an increased capacity for adaptive variation in changing environments due to sexual reproduction. Sex allows beneficial mutations to be passed on to progeny, even though the breakup of advantageous mutations is likely. In prokaryotes, many beneficial mutations are likely to be weeded out in the competition that arises in clonal interference, which is when mutations in different clonal lineages of prokaryotes compete for fixation in the population (Colegrave and Collins 2008). For these reasons, asexual organisms are usually considered to be *less* evolvable than sexual ones (e.g., Williams 1966), despite the evolutionary persistence of ancient asexual lineages even in animals. However, the persistence of those lineages, plus experimental evidence that tells only slightly in favour of sexuality for increasing adaptation (de Visser and Elena 2007), mean that existing theories about the evolutionary advantages of sexual reproduction have considerable gaps. They are premised on a rather limited account of eukaryotic sex (see O'Malley et al.

2013). There is some evidence also that sexual reproduction reduces variation even as it prolongs lineage identity (Gorelick and Heng 2010), and this does not fit squarely into standard discussions of evolvability.

But mechanisms aside, a big conceptual problem for evolvability is that it invokes a long-range kind of fitness – not 'myopic' fitness to do with adaptation to existing and imminent environments, but fitness much farther along the evolutionary road (Brookfield 2001). Tied to this problematic issue is the question of the selection of evolvability. Evolvability is argued to be an *explanation* of phenomena such as hypermutation and LGT, because evolvability itself has been selected (e.g., Earl and Deem 2004). The controversy about hypermutability, for example, revolves around how directly the associated evolvability has been selected (Pigliucci 2008). In many cases, reported instances of the selection of evolvability can be explained as a by-product of selection for something else. In other words, the mechanisms of so-called evolvability might have been selected for one reason or another but not for evolvability itself (Kirschner and Gerhardt 1998).

Population geneticist Michael Lynch (2007: 8603) identifies four of the 'buzzwords' of recent evolutionary biology as 'complexity, modularity, evolvability and robustness'. They are also popular topics in philosophy of biology. All of them are usually discussed in adaptationist terms even though a considerable body of evidence points towards non-adaptive explanations. Mutators, the bacteria exhibiting hypermutable capacities in certain conditions, are very commonly interpreted as having adaptive benefits for the population in which they occur, and thus as contributing to the group's evolvability. But mutator populations in long-term experiments show only very slight fitness gains over non-mutator populations (Elena and Lenski 2003). A different problem for the adaptiveness of LGT, as noted above, is that although transfers can be selected for at the level of the individual organism, they might ultimately result in the extinction of the population (Rankin et al. 2010). As Lynch repeatedly points out, 'it is by no means clear that an enhanced ability to evolve is generally advantageous' (2007: 8603). Unless the notion of evolvability is tendentiously linked to the notion that increased capacity for generating variation will be always or even mostly adaptive, then all that can be claimed is that different lineages exhibit different rates of genetic variation and that at some historical points rates take off and at others not, with a variety of evolutionary consequences.

But given the discussion of the preceding section, it may not make a great deal of sense to separate out macrobial and microbial evolvability; instead, their collaborative interactions might be more effectively conceived as the appropriate level at which evolvability should be understood. Philosopher Kim Sterelny (2004) uses symbiotic systems such as *Wolbachia* and insects, or molluscs and sulphide-oxidizing bacteria, to outline a mode of inheritance that is 'sample-based' rather than 'information-based' (that is, strictly genetic). Sample-based systems are highly evolvable, he suggests, because of the major shifts in phenotypic capacities produced by such liaisons. In a different comparison of the evolution of multicellularity in Volvocaceans (plastid-endowed cells that have formed lineages of unicellular, multicellular and in-between forms), Sterelny (2006b) discusses the 'arrested' multicellularity of *Volvox* lineages in contrast to the full-blown multicellular diversity of the metazoans. According to his analysis, the metazoans have been the most evolvable. Evolvability, he concludes, can be understood as a lineage property – a disposition to evolve over time (Sterelny 2006b: 186). This claim and others like it obviously invoke clade selection (that is, of species or larger evolutionary groups), which is still much debated in the evolutionary literature. But even if clade selection could be agreed upon, extinct clades cannot summarily be diagnosed as 'less evolvable' than extant ones (Brookfield 2001). Many more details are required in regard to short- and long-term fitness and the underlying bases of those fitnesses, meaning that any well supported claim about evolvability is likely to be highly particular and retrospective (rather than making claims about future evolutionary potential).

But more importantly, thinking of multilineal groups as adaptive, evolutionarily entrenched units, and following Sterelny's idea of evolvability as the symbiotic achievement of new phenotypic capacities, means focusing instead on the ability to form collaborative groups. In this view, the simple comparison of evolved capacities does become relevant. Metazoans have evolved diverse morphological structures but limited metabolic capacities; the opposite situation obtains for prokaryotes. At the biochemical level, prokaryotes are massively more capable than metabolically conservative eukaryotes. The former interact with a far greater range of environments, whereas the latter must maintain greater environmental consistency within and across generations (Poole et al. 2003). Now, while morphology allows certain kinds of adaptation – including the ability to incorporate other cells – metabolic flexibility can completely transform phenotypes, environmental requirements and the developmental processes of all the organisms involved. When new metabolic capacities with major survival and reproductive benefits are conferred in mutualisms, it would be odd not to think of this as evolvability, and unhelpful to try and separate microbial and macrobial capacities for co-producing it. A straightforward claim to make, therefore, would be that evolvability of an unmysterious sort is very commonly an outcome of sustained multilineal interaction. The adaptiveness of single organisms and their progeny is enhanced by such collaboration, and the tendency to contribute to cooperative inter- actions can become entrenched over evolutionary time (not always for adaptive reasons). The eco-microbiological perspective of Chapter 5 will make this point in more detail.

Whether evolvability is understood as selected in its own right or simply as a general propensity of evolving collectives, it is undeniable that pro- karyotes make good model systems for its study and for the study of any evolutionary process, including the more abstracted ones Lynch lists as buzzwords. Since microbial groups evolve in 'real' time (that is, human- observed time), contemporary experimental evolution can capture the effects of all the numerous processes that produce adaptive and non- adaptive phenomena in living systems. Microorganisms are therefore likely to be the model systems on which the best studies of evolvability and associated properties can be conducted. Ergo, they are probably the best model systems for evolution all round, runs this argument, and this is the second main theme of the book as a whole. It will be pursued further in the following chapters.

Concluding implications for philosophy of biology

[T]he theory of Natural Selection ... implies no *necessary* tendency to progression. A monad, if no deviation in its structure profitable to it under its *excessively simple* conditions of life occurred, might remain unaltered from long before the Silurian Age to the present day. I grant there will generally be a tendency to advance in complexity of organisation, though in beings fitted for very simple conditions it would be slight and slow. How could a complex organisation profit a monad? If it did not profit it there would be no advance (Darwin 1887 [1859]: 210).

Darwin himself felt compelled to argue for a general theory of evolution that could place unicellular life (monads) on an equal footing with multicellular organisms. Although some of the answers to the question of how morpho- logically complex organisms have evolved need contingent historical explan- ations, selection – along with other general evolutionary processes – is still important for an understanding of these historical events. Darwin's answer to the question about why microorganisms are not more complex is embedded within his argument about the universality of evolution for all organisms. The material in this chapter indicates that mode and tempo of microbial evolution, and even just prokaryote evolution, are not utterly distinct from the evolution of multicellular organisms. This conclusion is, however, a little more complicated than saying 'no difference'. No special account of evolution need be given exclusively for microbes, but any general theory of evolution needs to accommodate the greater range of evolutionarily relevant activities in which microbes participate. Perhaps this is also the place to make a plea for other organisms that have so far been invisible to the philosophy of biology, such as the enigmatic fungi. Although more closely related to animals than any other major clade, fungi have not featured prominently in either mainstream evolutionary theory or philosophical analyses of such theory. Looking at gaps and neglect is not an exercise for its own sake, but a constructive project: to work out how the current framework stands when confronted with apparently anomalous and wayward organisms.

Although I have already noted that some evolutionary microbiologists do indeed call for dual prokaryotic and eukaryotic perspectives on evolutionary processes and patterns, in the various aspects of evolution I have discussed, making a qualitative distinction between the two seems hard to justify. Different emphases may certainly be warranted, as, for example, when social evolutionary theory is used to understand microbial evolution. Rather than the ability to discriminate kin being a mechanism for kin selection as it is in the animal world, dispersal restrictions (keeping kin close) may be more relevant to the social evolution of prokaryotes (West et al. 2007a; however, cf. Verbruggen et al. 2012). A great wealth of experimental evolution on microbes is now finding and theorizing these differences, and then investigating whether and how they exist in the multicellular world. It is already clear that general social accounts of evolution not only can but *do* include microbes, and these social theories of evolution are flourishing because they can do so.

Numerous dimensions of LGT and endosymbiosis need further exploration in order to contribute to a fuller understanding of evolutionary processes. These dimensions need to be understood not just for prokaryotes, however, but for all life. Rare as it may be for animals and plants to acquire functional genetic material laterally, they nevertheless do so, and some lineage-forming events, such as hybrid speciation, or the incorporation of retroviruses (for example, at the base of the tree of mammals, when retroviruses endowed the ancestors of mammals with the ability to form a placenta), have conceptual and phylogenetic implications that are similar to those of LGT and endosymbiosis in microorganisms. The greater frequency of such events in the evolution of prokaryotes, and the potential selection of such mechanisms and their particular results, do not mean that making major distinctions between prokaryote and eukaryote evolution (or microbial from macrobial) is a compelling option. Again, one of the main messages from a consideration of LGT and endosymbiosis is simply that traditional accounts of plant and animal evolution need to be more inclusive, and that it may be most effective to construct any properly general account of evolution from the microbes up, rather than the macrobes down.

To reinforce this point, let me go back to the discussion about the units of selection vis-à-vis microbial evolution. Microbes of course can and often need to be understood as individual organisms and single-lineage groups. But much of the material above shows how a great deal more can be understood about evolution by looking at multilineage groups in which microbes play such a range of roles. Whatever concepts of group and group selection are deployed, microbes can illuminate them and provide an extraordinary range of examples and counter-examples for any theoretical account of levels of selection. Perhaps more telling than short-term accounts of group fitness, and their specifications for how cooperation is maintained, are the persistent evolutionary collaborations between macro- and microorganisms that have lasted hundreds of millions of years – underlying, for example, major evolutionary events such as the transition to terrestrial lifestyles in plants. Rather than focusing on selfish genes or arguing against the very notion as philosophers do, a 'selfish metabolism' account of evolution might be more appropriate. This account is one in which evolution is driven by whatever organismal combinations it takes to achieve 'the thermodynamically feasible conquest of the available chemical space' (de Lorenzo et al. 2013: 10). The 'selfish' is only there to provide a metaphorical jumping-off point, however: metabolism always has to be conceived interactively, whether in individual or multiple organisms.

It should not be surprising that a major message of this chapter is such a bland one: that evolution has to be understood across the spectrum of biodiversity. All organisms exist in fluctuating environments, multicellular no less than unicellular, and all living things and the associated genetic elements permeating life (for example, viruses, plasmids and so forth) evolve or become extinct. The important issue is to find a way to connect, rather than separate, the diversity of experimental and observational data with available theoretical frameworks, so that a better picture of evolution can be drawn. Indeed, many calls for a broadened modern synthesis do just this, and while some of them may not particularly emphasize microbial evolution, others certainly do (e.g., Koonin 2009; Rose and Oakley 2007). In addition to theoretical reasons, there are good practical reasons not to exclude microbes from a fully modern synthesis of evolutionary biology: because of their amenability to experimental studies. Not treating prokaryotes and other microorganisms as model evolutionary systems would be counterproductive, even if their size and biological differences made them seem just temporary surrogates for 'proper' zoological models (see Chapter 6).

One framework for a broader synthesis of evolution might be found in a continuum, with 'simple' microbes at one end and 'complex' organisms such as ourselves at the other. At the simple end, evolution occurs in a faster, messier way that involves the violation of established boundaries between evolving biological individuals and a range of processes that stretch contemporary neo-Darwinian concepts. At the other end, these boundaries are maintained, the tempo of evolution is more gradual and fewer conceptual changes are needed. But biology itself is not so linear, and smallness is not synonymous with simplicity. As the discussions above about multilineal groups showed, big things can make huge adaptive leaps when partnered by tiny microbes. The very smallness of microbial life means it is part of all other life forms and this fact – plus multiple exceptions to the continuum schema (see O'Malley 2010a) – discourages its adoption as a framework for how evolution works across biodiversity. Instead, a more integrative, interactionist model of evolution is required in which traditional macrobial evolution is combined with major insights from microbial evolutionary research.

This chapter has at least put on the philosophical table a number of reasons why long-term interactions between microbes and other organisms add new dimensions to existing evolutionary frameworks. As the evo-devo and earlier modern synthesis cases show, making universalized claims about evolution but then restricting them to particular organisms is a self-defeating strategy. It makes more sense to work 'upwards', in the sense of small to large: find your generalizations in the most common and pervasive life forms and then work out where to put your occasional anomalies such as animals. An ecological perspective is central to understanding evolution from any organismal basis, and the next chapters will elaborate on microbial ecology in a way that reinforces these conclusions about evolution.

CHAPTER 5

Microbial ecology from a philosophical perspective

Microbial ecology until recently was a minor stream of microbiology. Molecular innovations have now pushed the field to the forefront of microbiological research. At the same time, these methods have generated a whole raft of conceptual issues about microbial biodiversity. These are not simply questions of how many microbial species there are, but more basic ones about whether taxa are the entities that should be the units of analysis or whether functionally interacting groups are more appropriate. Communities and ecosystems are increasingly the object of attention in microbial ecology. Analyses at these levels raise further questions about whether concepts of organism, multicellularity and reproduction should be understood in dependence on animal-based notions. Likewise, models of the distribution of microbial biodiversity and community assembly have been borrowed from large-organism ecology, but their application in microbial ecology suggests new approaches to macrobial ecology.

A brief history of microbial ecology

Most histories of microbiology do not emphasize ecological perspectives. This is at least in part because of the emphasis on laboratory-based pure culture methods and the medical applications of microbiology. However, in the early days of microbiology, when taxonomic efforts dominated, many systematists had no choice but to gather their organisms from environmental sites and even study them there, since laboratory cultures did not exist. For example, Ehrenberg – the taxonomist mentioned in Chapter 3 – collected or was sent samples from the Red Sea, the Rocky Mountains in the United States and a great variety of other aquatic and terrestrial sites (Ehrenberg 1838; O'Malley 2007). Later, as the pure culture method rose to prominence, a few microbiologists augmented their laboratory work with techniques that approximated environmental conditions. They also laid more explicit foundations for microbial ecology. These researchers argued that making 'controlled' approximations of natural environments was a rigorous scientific strategy that would deliver more knowledge than pure culture could (Stanier 1951).

The pure culture method and its importance for establishing microbiology as a science has already been discussed in Chapter 3. However, valuable as the laboratory study of microbes was for understanding what they did to humans, many bacteriologists and other microbiologists in the early twentieth century still felt that microorganisms were worthy of study for their own sake (e.g., Brown 1932). In addition, despite the general acceptance of bacterial monomorphism (a single fixed form), numerous microbiologists were concerned that unnatural conditions prevented true understanding of more variable bacterial physiologies and interactions. As a result, even though medical microbiology dominated most approaches to the study of microbes, the broadly environmental study of microbial biodiversity persisted. There are a number of iconic figures associated with this persistence and they are all considered to have made seminal contributions to the emergence of microbial ecology as a field (Figure 5.1). Soil and marine microbiology play particularly strong roles in this story. Although these microbiologists are by no means chosen via rigorous historical analysis, they are identified by practitioners of microbial ecology as leading actors in the development of an approach that went beyond pure culture.

The scientist who is usually given precedence in this historical narrative is Sergei Winogradsky (1856–1953), a Russian soil microbiologist.[1] Findings of nitrifying and sulphur-oxidizing bacteria led him to propose a model for bacterial chemolithotrophy and chemoautotrophy (Winogradsky 1949; Glossary). Winogradsky unconventionally named his discoveries of bacteria based on their biochemical capacities and not their shape, despite the latter being standard procedure at the time (Dworkin 2012). He made these discoveries with his 'elective culture' technique, now called enrichment culture, with which he designed growth media that would sustain only very specific (elective) modes of nutrition. Winogradsky inoculated the media with organisms from natural sources, rather than resorting to the increasingly common practice of using laboratory stock (Dworkin 2012). Although he worked in the laboratory with manipulated and simplified environments, Winogradsky's emphasis on constructing cultured systems based on natural conditions has often earned him the distinction of being the first microbial ecologist (e.g., Stanier 1951).

---

[1] Winogradsky did his notable early work in Germany and Switzerland, went back to Russia, then became an exile researcher in France due to the Bolshevik revolution.

Figure 5.1: Some major historical figures in microbial ecology: Winogradsky, Beijerinck, Baas Becking, Waksman, van Niel, ZoBell, Jannasch. Baas Becking image courtesy of *Biografisch Woordenboek van Nederland*; van Niel used with permission from Edward Weston/Viscopy; ZoBell used with permission from Scripps Institute of Oceanography Archives, University of California San Diego library; Jannasch used with permission of Woods Hole Oceanographic Institute.

The Dutch microbiologist mentioned in Chapter 3, Martinus Beijerinck, employed the same approach. He was employed in both industry and academia, and in 1895 took up the first chair in microbiology at what is now Delft University of Technology. Beijerinck studied microbes eclectically – not rigorously enough as far as Winogradsky was concerned – and made extensive discoveries about the metabolic capabilities of bacteria, yeast, algae and protists (van Iterson et al. 1983 [1940]). He also identified the tobacco mosaic virus and was first to realize that viruses and bacteria were different due to the former entities needing a cell to host them (see Chapter 3). Beijerinck made these findings through use of the elective culture method, which he called the 'accumulation method' – possibly to appear to be doing something different from Winogradsky (Ackert 2006). Conceptually and theoretically, Beijerinck pursued an evolutionary and genetic agenda well in advance of his peers. He speculated that genes – theoretical entities at the time – were inheritable enzymes, and he 'rediscovered' Mendel's work on genetics before his contemporaries. He viewed his ecological work within a unified theory of inheritance that included multicellular and unicellular life, and he anticipated that bacteria would become experimental systems for genetic research (Beijerinck 1900–1901; Theunissen 1996; see Chapter 6). In 1913 he coined the word 'mikrooekologie' – Haeckel having already come up with 'Ökologie' in 1866 – to describe the ideal environmental approach to microbial study (Beijerinck 1913; de Wit and Bouvier 2006). His place in the history of microbiology, and especially microbial ecology, was assured when the label of 'founder' of the Delft School of Microbiology was retrospectively bestowed on him (van Niel 1949; Robertson 2003).

Beijerinck's successor at Delft was Albert Kluyver (1888–1956), whose contributions to comparative and ecological biochemistry are justly celebrated. But it was Kluyver's student, Cornelis van Niel (1897–1985), who exported the Delft School of microbiology and its ecological perspective to North America and thus launched the larger tradition of microbial ecology outside Europe. Van Niel was already highly esteemed for his work on purple- and green-sulphur photosynthesizing bacteria, and his research had established a unified understanding of oxygenic and anoxygenic photosynthesis in plants and bacteria. Convinced that pure culture was inadequate for the proper physiological study of microbes, van Niel advocated a much broader view of microbiology, which he called 'general microbiology'. He believed microbiology had to encompass the evolutionary forces that operated on bacteria (van Niel 1955; 1966; Spath 2004). Microbiology in its more medically oriented practice had not at that time – or even later – found it useful to take up evolutionary theory.

Although a critic of pure culture, van Niel also acknowledged the limitations of elective culturing, noting that the environments created by this method were unnaturally simple and sometimes contaminated (van Niel 1955). He was interested in developing a unifying perspective for non-molecular as well as molecular studies of microbes so that microbial interactions could be understood (van Niel 1966). Van Niel emphasized that there are many circumstances in which 'environmental factors rather than genetic factors influence directly the enzymatic composition of microbes' (1949: 172), even though he was well aware of molecular biologists asserting that genes 'controlled' enzymes. The one-gene-one-enzyme hypothesis would probably not hold in light of future knowledge of environmental factors, he thought. Van Niel's endorsement of an environmental perspective had a lasting impact, not least because of his influence on an important cadre of younger microbiologists as well as some up-and-coming molecular biologists. Many of them were students on van Niel's famous general microbiology course (Spath 2004). Several of these graduates developed specific fields of microbial ecology in significant ways (see below) as they advanced van Niel's vision of microbiology.

Working in roughly the same era as van Niel was a famous soil microbiologist, Selman Waksman (1888–1973), who also explicitly acknowledged Beijerinck and Winogradsky as his intellectual forebears. It was not by chance that soil microbiology played such a large role in the development of microbial ecology. Although it was based in applied agricultural science, microbes in soil were not seen solely as pathogens to be eliminated, as they were in medical microbiology, but as resources that could make soils more productive. Waksman's work, though based in an ecological approach, was in important respects absorbed into the medical microbiology tradition with his discovery of antibiotics, such as streptomycin and neomycin (Schatz et al. 1944). He was awarded a Nobel prize for these discoveries, although there was controversy about how much of the discovery work was done by his research assistant and PhD student, Albert Schatz (Schatz 1993; Waksman 1953). Despite these medically related accomplishments, Waksman's *Principles of Soil Microbiology* (1927) was a formative text for microbiologists with an ecological bent.

Another exemplar of the Dutch approach to microbial ecology was Lourens G. M. Baas Becking (1895–1963), who also moved from the Netherlands to the US (temporarily).[2] His speciality was salt lakes and the organisms that lived in them, some of which were halophilic prokaryotes. He made copious observational and experimental studies of both eukaryotes and prokaryotes in those environments, and many of his findings are still sources of insight (Oren 2011). Baas Becking took a broad view of how microbiology should be understood and made efforts to expand its conceptual toolkit. He employed the term 'geobiologie' to describe research that sought to understand organisms in relation to their geological and chemical environments (1934, in de Wit and Bouvier 2006). Well before the 1970s 'Gaia' revival, Baas Becking used the word to refer to the relationships between all organisms and the total geobiology they inhabited (1931, in Quispel 1998). His Gaia thoughts were an extension of his interest in symbiosis, in which he included not just positive interactions between dependent organisms but any enduring relationship of mutualism, commensalism or parasitism (1938, in Quispel 1998; for a history of other views on symbiosis see Sapp 2004).

The microbiologists who took microbial ecology closer to the mainstream – in part because they developed ways in which to culture previously

---

[2] Baas Becking studied in Utrecht, went to California in the 1920s, and then came back to the Netherlands, to Leiden, for the 1930s. After pre- and post-war years in Java, he spent the last dozen years of his career in Australia where the Baas Becking Geobiology Laboratory was established in his honour from 1968 to 1985.

unknown organisms – were direct and indirect intellectual descendants of van Niel. One of the former was Robert Hungate (1906–2004), who was an early rumen ecologist. Hungate studied the microbial occupants of the first stomach chamber for cellulose digestion in ruminants such as cattle. He was one of van Niel's students – the only student, in fact, in van Niel's first and subsequently hugely popular general microbiology course at Stanford University's Hopkins Marine Station (Hungate 1979). Hungate worked out ways to calculate in situ the metabolic activity (especially methane production) of the anaerobic microorganisms in cattle rumen, and devised methods by which to culture them with habitat simulations in vitro (Hungate 1960). He designed and constructed several innovative pieces of equipment to carry out his research, and made major advances in the understanding of complex ecosystems formed by multiple organisms and numerous biochemical pathways. Hungate was adamant that understanding what went on in any microbial ecosystem such as the rumen involved 'quantitative measurement of the entire complex as well as its individual components' (Hungate 1960: 354). His broader contribution to microbial ecology is understood in terms of 'anaerobic microbiology', and his techniques used successfully for obligate anaerobes in many environments.

Two more technological innovators made enormous contributions around the same time to marine microbiology. Claude ZoBell (1904–1989) is commonly called the 'father' of his subfield, which he helped establish with his book, *Marine Microbiology* (ZoBell 1946). Although marine bacteria had been studied in a limited way since the nineteenth century, and marine protists a little more so, ZoBell developed effective methods by which to sample, examine microscopically and culture marine bacteria, especially those from the ocean depths. He coined the term 'barophiles' for organisms that dwelt obligately at high hydrostatic pressures and low temperatures (ZoBell and Johnson 1949). ZoBell also made major contributions to early 'petroleum microbiology' with his studies of the microorganisms in deep-sea oil deposits (Stone and ZoBell 1952). He took out patents on microbial processes that he thought would increase oil flow in subterranean rock formations (see Bryant 1987 for why they couldn't work). His research drew attention to biofilms, then called 'attached films', which have become a major focus of today's microbial ecology.

Holger Jannasch (1927–1998), another marine microbiologist of the deep ocean, saw his life's work as deeply influenced by van Niel (Jannasch 1997). Jannasch pioneered the microbiological study of ocean-floor thermal vents and devised numerous techniques and tools for culturing microbes in situ and bringing barophiles alive to the sea surface. A famous story is told of how he developed one of these technologies after the Woods Hole Oceanographic Institute submersible sank in 1969. With it sank the full lunchboxes of the submersible's operators, who survived. But when the submarine was brought back to the surface a year later, Jannasch found that the lunches were soggy but not decomposed. He was inspired to send down a lunchbox-like piece of homemade equipment to the ocean depths in order to inoculate and culture it with the bacteria living there (Jannasch and Taylor 1984). Jannasch also learned from the lunchbox how very slow microbial metabolism was in such conditions, in contrast to the high productivity at deep-sea hydrothermal vents, where microorganisms could often symbiotically sustain far larger organisms.

Near the end of his career, Jannasch witnessed and partly participated in the early use of genomic and molecular phylogenetic tools in microbial ecology. He was worried that the community might be 'going overboard' with these techniques, due to the deprivation of data that systematists had experienced prior to the emergence of these methods. Don't forget, he argued,

> Microbial ecology, the interaction of the species with the environment and one another, is process-based, and the phylogenetic positioning of a phylotype can only give hints of the organism's actual function in the environment. Discarding pure culture approaches ... or using the term unculturable microbes means throwing the baby out with the bathwater (Jannasch 1997: 36–37).

We will come back to this issue later in the chapter.

Marine microbial ecology at this point began to make closer connections to a broader field of aquatic microbial ecology, the promotion of which was taken up in the work of people such as E. J. Ferguson Wood (1904–1972). Wood was greatly influenced by Baas Becking, and the many textbooks Wood wrote on the general topic helped this area of ecology to grow considerably in subsequent decades (e.g., Wood 1965; 1975; Psenner et al. 2008). This was despite a book reviewer complaining that Wood's account of marine microbiology should not include phytoplankton, which – although unicellular – were 'the subject of other branches of science' (Sorokin 1971: 515). Disciplinary boundaries did not in the 1960s and 1970s offer great assistance to the development of microbial ecology. It would be more accurate to say they actively obstructed the field's emergence (Susan Spath 2013, personal communication). Wood rose above those circumstances to articulate a broader vision of the field, in which he tried to follow Baas Becking's example by emphasizing the symbiotic nature of life: 'It will be realized that life in the sea is a vast symbiosis or metabiosis and that no organism can exist independently of others' (Wood 1958: 9).

Marine microbial ecology has had considerable impact on large-organism marine ecology, particularly because of the incorporation of microbial contributions to marine food webs (Pomeroy 1974; Azam et al. 1983). Photosynthesizing bacteria and carbon-consuming microorganisms make major differences to understanding carbon cycles in the ocean. Despite their small size, microorganisms comprise ten times the biomass of multicellular organisms in marine ecosystems (Pomeroy et al. 2007). Viruses are also recognized now as having large effects on these food webs as they lyse cells and change population structures and carbon availability (Fuhrman 1999; Suttle 2007).

Baas Becking and other halophile researchers had laid the foundation for a focus on unusual or extreme environments in microbiology, as did the 'barophily' on which ZoBell and Jannasch concentrated. Environmental microbiologist Thomas Brock took up this tradition and made major advances, while at the same time unifying 'microbial ecology' as a field rather than as separate areas of microbiological research in different environments. Brock (1966) emphasized along Delft School lines that ecology is physiology studied in the organism's actual habitat. He focused on hot springs, particularly those in Yellowstone National Park, and managed to sample several thermophilic (heat-loving) organisms, some of which were subsequently recognized as archaea (Stahl et al. 1985). Averse to the enrichment culture method because of its 'artificiality' (Brock 1995), Brock developed an 'immersion slide' technique whereby glass slides were lowered into hot springs and left there for microorganisms to grow on (similar to what Jannasch did in deep-sea hydrothermal vents). But it was through conventional isolation and culturing techniques that he made the discovery of *Thermus aquaticus*, a thermophile that grew best at 70 degrees Celsius (Brock 1967). The protein that allowed it to flourish in such high temperatures was subsequently commercialized as *Taq* polymerase, which is used for the DNA amplification technique of polymerase chain reaction (PCR).

It was at this point that thermophiles – including some organisms that flourished at even higher temperatures – became not only scientifically attractive biological curiosities but also commercial goldmines. As well as galvanizing industrial enzymology, the *Taq* polymerase discovery simultaneously gave microbial ecology a higher profile for academic and industrial research funding (Brock 1995). While continuing to develop thermophile research, Brock published a number of classic texts on microbial ecology (e.g., Brock 1966). He also wrote more extensive and now mainstream microbiology textbooks that in later editions included large chapters on ecology. Through this means he effectively 'ecologized' the teaching of microbiology. His approach, based on van Niel's, emphasized the integration of lab and field studies. But as Brock pointed out as late as 1987, much so-called microbial ecology still used pure culture methods or other ecologically 'discredited' methods, and continued to be segregated from other microbiological research (Brock 1987).

Despite these worries, there were many signs of progress and development in the field (Tiedje 1999). By the 1970s there were journals dedicated to microbial ecology: *Microbial Ecology* itself (first published in 1974) and *Applied and Environmental Microbiology* (which had started as *Applied Microbiology* in 1953 and broadened its scope in 1976). Also in the early 1970s, microbial ecology became the topic of international meetings. In 1977, the First International Symposium on Microbial Ecology (ISME) was held in Dunedin, NZ. Jim Tiedje, a prominent microbial ecologist, emphasizes the importance of this meeting for present-day microbial ecology (Tiedje 1999). He gives a large role to the creation of tools, materials and organizational approaches rather than theories. But at the same time, Tiedje warns against being 'infatuated' with molecular methods just as Jannasch had a little earlier. Tiedje's 'chronology' of important phases in microbial ecology selects microbial interactions in economically important hosts as the focus of the 1970s, mathematical tools in the 1980s, molecular tools in the 1990s, and for the 2000s, multiple methods and the era of community studies. In this current phase, predicts Tiedje, community-level patterns will be given mechanistic underpinnings. The very possibilities of doing so have come about not by counting organisms directly but by quantifying DNA and other sequences.

Molecular innovations in microbial ecology

Microbial ecology has not for the most part been included in mainstream ecology (see exceptions below and in Chapter 6), although both these ecological fields have been around for the same length of time – at least since the end of the nineteenth century. Although microbial ecology occasionally tried to integrate theory from plant and animal ecology, or to develop new models when macroorganismal ones are not easily applied, it has been largely a method- or tool-driven discipline, hampered by entrenched inabilities to quantify microbes – particularly prokaryotes and viruses – in their natural environments. In addition, the strong identification of much microbiology with medical microbiology, and its ongoing dominance by laboratory-based approaches, have meant that microbial ecology was only a minor strand of microbiology for much of the twentieth century.

That situation changed radically in recent decades as the overview above showed. First, from the 1970s onwards, field-based microbial ecologists followed Brock's lead in isolating potentially commercially valuable micro-organisms in extreme or unusual environments. Then, as Tiedje noted (1999), field-based approaches to microbial ecology experienced a massive influx of molecular tools from the 1980s onwards. Small-scale molecular approaches flourished in the 1980s and large-scale genome-based approaches in the 1990s and 2000s. Methods to analyse the molecular diversity of microbes in their environments were pioneered with 16S rRNA by Norman Pace and colleagues (e.g., Stahl et al. 1985). Novel sequences and completely unknown clades of organisms were found by this approach. For many practitioners, it was the ultimate means by which to fill in the entire tree of life (Pace 1997). Gene-sequencing techniques liberated microbial ecology from pure culture approaches by enabling the DNA-based identification of 'missing' microbes: organisms that seemed to exist under the microscope but could not be cultured, and organisms with metabolisms that should exist because of thermodynamic potentials but could never be observed under the microscope or in pure culture.

It was well known in marine microbiology in particular that 'the numbers of bacteria recorded in sea water do not seem sufficiently high for the performance of the reactions which are believed to be of bacterial origin' (Wood 1958: 6). ZoBell had noted that direct microscopic counts were unable to distinguish between dead and living cells, and in addition, that such counts said nothing about any organism's physiology (ZoBell 1946). Hungate, too, had expressed concerns that even after strenuous efforts only some of the rumen organisms could be cultured, identified and counted, meaning that little could be known of the total community composition let alone its metabolic activity (Hungate 1960). In the mid-1980s, the discrepancy between microscopically visible and culturally 'viable' cells in environmental samples was named 'the great plate count anomaly' (Staley and Konopka 1985). This labelling of the problem stimulated a fairly crude estimate that fewer than one per cent of the microbial cells in most environments could be cultured – both eukaryotic and prokaryotic (Caron et al. 1989; Amann et al. 1995; see Chapter 3). Methods to detect more diversity flourished when molecular methods became more efficient in the late 1980s with the advent of PCR amplification of DNA. Findings proliferated of completely unknown organisms, all inferred from 16S rRNA sequences. However, although ribosomal gene sequencing was of enormous value in indicating the scope of microbial diversity, it was still very limited in what it could say about the functions of these organisms in their environments.

Often seen as a solution, methods that focused on whole-genome sequences rather than single genes began to sweep through ecological microbiology at that same time they became essential to molecular phylogeny. Phage genomes had already been sequenced in their entirety in the 1970s, followed by complete microbial (prokaryote and eukaryote) genomes in the 1990s. This genome-wide investigation of molecular diversity at the individual genome level was then applied beyond domesticated microbes in laboratories to 'wild' ones as they interacted in natural environments. But this was a dual expansion, because it also went beyond individual organisms to encompass whole microbial communities. Environmental genomics or metagenomics (already mentioned in Chapter 3) are the labels given to this method of sequencing the DNA found in a 'natural' sample (Handelsman et al. 1998). Metagenomics, the most common name now, is based on the core tenet that microbes and molecular activities have to be understood in their ecological contexts, rather than as isolated entities in specially created laboratory environments. It thus can be seen as the culmination of the Delft School approach as envisaged by van Niel and those he influenced.

Molecular techniques in microbiology have been made to stand in for the sorts of field studies that plant and animal ecologists are able to perform. Previous chapters already showed how single-organism genome sequences led to radical revisions in evolutionary understandings of microbes and microbial classification, especially as sequence analyses indicated the extent and importance of shared genetic resources in communities. While metagenomic studies often use DNA sequence merely to quantify biodiversity (with many caveats because of the ongoing methodological implications confronting both data gathering and analysis)[3],

---

[3] Methodologically, there are numerous weaknesses that go beyond well-recognized sequencing issues, especially to do with the inadequacy of comparative samples (replicates) and the inability to generalize reliably from single samples (Prosser 2010). Other challenges include the overestimation of genes and organisms in many samples (Gomez-Alvarez et al. 2009). An additional basic problem is that some of the DNA is from dead cells, and indistinguishable from the DNA of living cells.

the latest wave of molecular microbial ecology is most concerned with metabolic, signalling and regulatory interactions in natural environments. In this strand of research, the emphasis is on the functional roles of genes in environments, and the dynamics of how such genes are used in a range of mixed-species communities over ecological and evolutionary timescales. Some of this work looks at metabolic interactions between different groups of organisms; other research is more interested in examining the molecular relationships between groups of organisms and environmental geochemistry.

The transformation of knowledge in any of these areas is still far from complete. To go further, an even wider array of methods is being developed to work in concert with metagenomic techniques, which them-selves may be focused to produce more 'targeted' outcomes (that is, focusing on specific sequences and not the full communal genetic resource). Results produced by these diverse methods are not always easy to integrate, especially when new culturing techniques, for example, produce different biodiversity estimates from metagenomic ones (Shade et al. 2012). But methodological discord aside, metagenomic analyses have already produced numerous correlations between genes and community function. Most of these observations are not at present captured by mechanistic models explaining biochemical activity over different time-scales. However, even with an emphasis so far on description and correl-ation, metagenomic and other microbial ecological findings have not simply filled in the templates handed down from plant and animal ecology. New conceptual frameworks are being explored and developed to encom-pass the flux and plasticity of microbial ecosystems. For example, rather than taking as fixed an organism's inheritance and ecological role, metage-nomic analyses have emphasized the flexible, collaborative and network-based nature of microbial life. Consequently, ecological approaches have transformed not only the ways in which microbiology is conducted, but also its basic notions of the fundamental units of ecology, evolution and life.

Biodiversity and the units of analysis in microbial ecology

Ecological research carves the biotic and abiotic world up into many sorts of units. Biodiversity is the basic platform of knowledge on which the broader projects of ecology and conservation biology are built, as well as evolutionary understanding.[4] In order to grasp what is going on in any particular environment, something has to be known about which are the entities occupy that location. In plant and animal ecology, this usually involves sampling and counting individual organisms, although sometimes counts can be exhaustive. In microbial ecology, counting has always been a problem (see above) and sampling even more so (see below). The entomologist and evolutionary-ecological theorist, Edward (E. O.) Wilson, in his treatise on biological diversity in the early 1990s, claimed that 'bacteria await biologists as the black hole of taxonomy' and thus biodiversity research was obstructed (1992: 148). Although he thought that few scientists cared or even dreamed about making inroads into bacterial biodiversity, the history of microbial classification makes clear this is not the case (Chapter 3). And as Wilson himself enthused, microbial ecology is what he would do if he could make his scientific choices all over again: there is so much to learn, so many new tools and a sense of great scientific excitement because 'the smaller the organism, the broader the frontier and the deeper the unmapped terrain' (Wilson 1994: 364).

But many microbial ecologists and taxonomists, who are at least as passionate as Wilson about their field, also admit the daunting task they face in identifying and characterizing microbial biodiversity. For example, even now in the age of a deluge of molecular data, small and few samples mean that the very low-abundance taxa in communities such as deep-sea hydrothermal vents repeatedly escape detection (Sogin et al. 2006). Follow-up studies with more intensive sampling of the same environment have also proved unable to resolve the community structure of the organisms in that habitat, even at fairly coarse levels (Huber et al. 2007). Diversity is just not tailing off yet with current methods. These low-abundance organisms may be biologically important and evolutionarily ancient, so methods need to be devised to ensure their representation in environmental samples and to test their functional importance. In classic ecological research, species are the basic units of biodiversity. Discussions of biodiversity units and measures are still at a formative stage in microbial ecology, particularly in regard to how species richness and relative abundance are calculated.

---

[4] One of the pressing problems in macroorganismal ecology is to decide which units of biodiversity are the targets of conservation strategy (Faith 2007). For reasons that will become clear below, I will discuss biodiversity units primarily in relation to ecological analysis rather than their appropriateness for conservation biology. However, some inevitable conservation implications will be explored in the concluding chapter.

Species and OTUs

There is a tendency to take species abundance as the prime indicator of biodiversity despite the many problems that accompany this strategy, whether in large-organism or microbial ecology. Many standard biodiversity counts, which aim to achieve a total understanding of diversity in a particular environment, exclude bacteria and archaea, 'because the species concept used for eukaryotes cannot be applied to these two taxa' (Appeltans et al. 2012: 2191). Brock called the very notion of species biodiversity in microbial ecology an 'infamous subject' (1987: 6). No matter which definition was used, he believed, it would get the diversity of any ecosystem wrong. This occurs in part because many standard species definitions are too broad (for example, phylotypes – the entities designated by 16S molecules) and they underrepresent biodiversity in any location. Species are also a problem because designating and classifying lineages of organisms via molecular proxies will not often capture the aspects of biodiversity that are most crucial to ecosystem function. Even though early efforts to classify environmental DNA argued that, 'In principle, phylogenetic placement can be interpreted in terms of physiology' (Stahl et al. 1985: 1383), in practice this has often been a misleading strategy to follow. Although some metabolic capacities can be predicted on the basis of 16S sequences (for example, in methanogens, cyanobacteria and some nitrifying and methanotrophic bacteria), it is rare for this to be the case (Ward 2005).

What are the alternatives for microbial ecology? Stanier thought that Winogradsky's and subsequent work had elaborated a new concept of '*ecologically-defined* microbial species' (Stanier 1951: 37). This way of understanding species was built on the insight that pure cultures produced artificially selected organisms that were more amenable to laboratory conditions than their undomesticated relatives were. This meant they were not 'natural' in the sense of being selected solely by existing environmental conditions, and that too little would be understood of real-world evolution through their study. Most experimental evolutionists think this problem is just one of idealization, which is inevitable in any type of research (see Chapter 6). Nevertheless, it might be the case that ecological methods in themselves can define how species have to be understood. The ecotype concept of species could fit here (see Chapter 3). Ecotypes are based on an ecologically oriented population-genetic definition as applied to molecular data. The problem is, as Chapter 3 already mentioned, this definition precludes recombination and requires using sequence data to detect 'ecological distinctiveness'. The selective sweeps that are theoretically responsible for divergence do not operate uniformly, and cohesiveness may be the result of recombination instead (Shapiro et al. 2012). Although the ecotype concept may be useful for understanding biodiversity in some organisms and habitats, it is not for others – especially when highly recombinogenic organisms are involved.

But regardless of conceptual problems, molecular microbial ecology tends to identify 'operational taxonomic units' (OTUs), usually in the form of phylotypes, which are more often than not proxies for species. In addition, as the discussion of classification in Chapter 3 showed, ecological diversity in microbes can be understood and analysed at higher and lower taxonomic levels, especially phyla and strains. But even if 16S sequences – or rather, fragments of them due to sequencing constraints – are seen as the ideal marker of the most basic biodiversity unit, biodiversity 'counts' are not enumerations of everything in an environment. Samples of these environments are often extremely restricted (Curtis 2006). Because of this, there are many problems in quantifying microbial biodiversity even at the 16S level.

OTU calculations of microbial species richness are often supplemented by measurements of the relative abundance (evenness) of those units.[5] A variety of methods can be used to make those estimates, each of which is borrowed from traditional ecology. Sample-based parameter fitting, frequency projection and curve extrapolation can all be used to determine the diversity of the sampling environment (Curtis et al. 2006; Robinson et al. 2010). No single approach is problem-free, so combinations of these approaches tend to be used to gain better insight into the microbial diversity of any given environment, and to compensate for sampling effects. New measures have been devised specifically to cope with the low-abundance taxa in many microbial communities (Li et al. 2012).

However, consensus has seldom been reached for the species richness of microbes in any environment, and not at all for global species richness. Calculating total prokaryote species diversity is obviously a task fraught with even more uncertainty than estimates for one environment, even though efforts are being made to develop tools that enable better comparative measurements of different environments. But all calculations are contested, and what continues to circulate are the older vaguer

---

[5] Simply counting taxa/OTUs in a sample can give a useful indication of alpha diversity (within-sample diversity), but a better understanding of community structure needs to estimate beta diversity or between-sample variation over time or environmental gradients. Beta diversity gives insight into how environmental and spatial variables influence diversity patterns (Hamady and Knight 2009).

estimates of global species richness. These estimates underpin broad claims that 'only one per cent of microbial diversity has been/can be cultured'. Even in total eukaryote species estimates there are many problems in devising acceptable calculations (e.g., Mora et al. 2011), with fungi and protists the least likely to be estimated adequately (Bass and Richards 2011). This is not just a matter of bad measurement or estimation. If there truly is no global species definition applicable to all organisms – and Chapter 3 showed how very likely that is the case – then the unit that works as the basis for calculations of the diversity of some groups of organisms will simply not work for others (Doolittle and Zhaxybayeva 2009).

But one reason for optimism is that identifying the quantities and proportions of sampled organisms is usually done as a preliminary task. The real purpose is to find out more about functional diversity in these samples (Little et al. 2008; Green et al. 2008). 'Function' is used in the Cummins causal-role sense (for philosophers of biology), and refers mostly to biochemical activities when discussed in microbial ecology. Biochemistry is obviously dependent on individual cells, and while there is considerable value in understanding individual-organism physiology, this focus is often insufficient to understand functional diversity in situ. Association studies can show which organisms co-occur with others, which relationships are most important in particular environments, and how environmental perturbations or fluctuations affect those functional relationships (Gilbert et al. 2012). One finding all microbial ecologists have made over the decades, whether using microscopy, culturing, physiological experimentation or molecular sequence analysis, is that microbes very commonly work together, frequently with or within much larger organisms (see Chapter 4). Because of this tendency to function in multilineal groups, many microbial ecologists have argued or at least assumed that the most important unit of microbial ecological research is the community (assemblages of organisms from different evolutionary groups). It can be understood as a unit of biodiversity, although its definition is even more problematic than that of species.

Communities and ecosystems

Communities and their conceptualization have a long tradition in plant and animal ecology. Amongst plant biologists, an early idea with much currency was that plant communities could be conceived of as superorganisms. Frederic Clements (1874–1945) is most associated with this notion, and particularly the argument that plant communities grow, mature and die in the same way individual organisms do (Clements 1916). He was roundly criticized by Henry Gleason (1882–1975), who thought that communities were constituted by the interactions of independent individuals, and that the community concept was merely a human construct rather than a reference to a biological entity (Gleason 1926). Although Gleason eventually won the debate, this did not happen fully until the 1950s with the development of quantitative gradient analysis (Odenbaugh 2007). In the meantime, Arthur Tansley (1871–1955) introduced the more inclusive idea of 'ecosystem' (1935), on top of an earlier compromise that plant communities should be studied as 'quasi-organisms'. I will come back to ecosystems below, but want to note that philosopher of ecology Jay Odenbaugh (2007) shows that the extreme views – 'there are no communities' versus 'communities are superorganisms' – do not represent adequately either historical conceptual claims or the newer research that has interrogated these different views empirically. Communities demonstrate different degrees of interaction between their constituent populations and thus form a continuum of organization.

In microbial ecology, there is very little concern about any historical conceptual baggage that may come with the term 'community'. The word is used prolifically and only occasionally queried. If it is questioned, the perceived issues are pragmatic, having to do with the difficulties of deciding where one community ends and another starts, and how far to pursue the interactions of any given community (Konopka 2006). In these respects, they are the same practical issues of community designation in plant and animal ecology. The question at the conceptual level, however, is whether metagenomics and microbial ecology more broadly are dealing with 'communities of genomes' (sequence data gathered together) or 'genomes of communities', with the latter implying an identifiable functional entity (Doolittle and Zhaxybayeva 2010: 111). The reason why this matters in microbial ecology is because the second concept is likely to have explanatory relevance and the first one is not: the former would simply need to be described.

Sterelny (2006a), using plant and animal communities as his exemplars, distinguishes between purely phenomenological communities and causal-system communities. The latter is what Gleason opposed and wanted replaced by communities as mere aggregates of individuals. Like Odenbaugh but with more variables, Sterelny proposes a 'dimensional' perspective on communities, in which they vary along three axes of collective causal properties, internal integration/regulation, and boundedness (Sterelny 2006a; Maclaurin and Sterelny 2008). Sterelny shows that communities – beyond how any particular one fits into this three-dimensional space – can be used to explain biological phenomena (2006a). The pursuit of this explanatory capacity is the way in which the concept of microbial community is used in today's microbial ecology. Although descriptions of communities are obviously relevant and important, there is a range of community-scale patterns that most microbial ecologists believe to be in urgent need of explanation.

Contemporary discussions of community in microbial ecology show an increasing emphasis on functional and not just compositional biodiversity. In microbial ecology, community-scale functional effects – the 'community phenotype' – are the focus, regardless of the conceptualization of community. Although community structure and boundaries are important, and often correlate with, for example, disease states in hosts, it is observable community-scale function that enables researchers to bypass strict definitions of community and not even worry too much about the cohesiveness and reproducibility of community composition (Zarraonaindia et al. 2013; see below). Even if community members are transient and rare, they can contribute significantly to community-scale functions, such as the digestion of certain nutrients and exclusion of other organisms. Furthermore, these traits are often deemed 'heritable', as I will discuss below. This community-phenotype view is consistent with the notion of communities as units of adaptation, (as discussed in Chapter 4) but can accommodate 'neutral' assembly processes, too (see below).

A very popular research theme in metagenomics by publication number and public interest revolves around microbiomes and microbiota, especially those in human bodies. 'Microbiota' refers to the collectivity of organisms inhabiting a particular niche; the 'microbiome' is the molecular basis of that collective organism.[6] The human gut microbiome in the large intestine has been an especially intensive research topic, both because of its long-recognized importance in health and accessibility to researchers via faecal samples. There are many trillions of these organisms in every human intestine; they weigh around a kilo in a large adult human.[7] Taxonomic assignment of genes from numerous samples indicates that there are at least 1000 different taxa in these communities, the great majority of which

---

[6] Sometimes people use microbiome to mean microbiota (the organisms), probably because 'biome' – a term coined by Clements again – normally means a biotic community in a particular habitat. However, in most uses of 'microbiome' (including this book), the 'ome' refers to 'omic' data extracted from the microbiota.

[7] Total microorganisms in and on the human body comprise somewhere between 1–3% of human body mass (NIH 2012).

are bacterial (Qin et al. 2010).[8] There are creakingly old estimates of ten gut microbe cells to every human cell (Savage 1977, based on even older calculations). This ratio is repeatedly cited and has yet to be contradicted by new data or calculations. The revised ratio of 150 microbial genes to every human gene has been calculated on the basis of predicted genes from a large number of gut microbiome samples (Qin et al. 2010).

However, more important than quantity is the functional importance of these microbiota and microbiomic components to the overall state of the human ecosystem. Microbes synthesize some of the vitamins and amino acids that humans need; they have important developmental effects on human digestive and immune systems; and they play what appear to be major causal roles in various disease states, including diabetes, malnutrition, inflammatory syndromes, various cancers and allergies (see Chapter 3 for a discussion of obesity links). Very little is yet known about the microbial mechanisms of disease or health states, and the correlations – while plentiful and often roughly reproduced in different studies – are so diverse that they make experimental testing of these findings a major undertaking. At present, much of the experimental work is done in germ-free mice that have been 'humanized' by human microbiota transfers and dietary regimes, although it is recognized that these transplants often adjust to match the host's physiology (Dantas et al. 2013).

Although there is considerable inter-individual variation between the gut microbiomes of human bodies (and even more in the 'virome' or viral equivalent), this variation is most probably explained by differences in how each person undergoes the original colonization process (Schloissnig et al. 2013). At least some of the colonizing microbes are 'inherited' from family members, meaning that family microbiota – while still very different – are more similar to one another than microbiota of non-family members (Fierer et al. 2012a). In utero, humans are largely uncolonized, and the assembly of the microbiota begins during the birth process. Once it begins, there is a pattern of succession events, in which major changes in community composition, relative abundance and function can be detected (Caporaso et al. 2011). These community-scale changes are precipitated by diet for the most part as well as age-related health status. This means that major biological changes in human bodies are accompanied or even preceded by microbiota community development (Lozupone et al. 2012).

---

[8] Much more limited work has so far been done on eukaryotic microbes in the human gut or other sites due to the larger size and more complex organization of eukaryote genomes (Gilbert and Dupont 2011).

Pregnancy, for example, is correlated with dramatic changes in microbial community structure, and this composition is also implicated in the reduced insulin sensitivity of advanced pregnancy that is characteristic of diabetes and obesity (Koren et al. 2012).

But exactly which way causality flows and the precise mechanisms of those changes are still unknown, although several experiments and a few models do indicate microbial causation and possible pathways (e.g., Vijay-Kumar et al. 2010; see Chapter 3). A ground-breaking study of malnutrition – for a change from obesity – shows very clearly that remedying a particularly severe syndrome called kwashiorkor cannot be achieved simply by manipulating the microbiome with improved diet (Smith et al. 2013). As medical microbiologist and metagenomicist David Relman notes in reference to this study, such findings are a disappointing reminder 'that isolated factors, such as individual microbes or even entire microbial communities, alone cannot explain complex phenomena such as undernutrition' (Relman 2013: 531). Regardless of the limited experimental and explanatory support, therapeutic strategies – such as faecal microbiota transplants, and probiotic and antibiotic treatments – are being implemented for a range of disorders that have some association with gut microbiota disturbance. These associations show signs of being mediated by the immune system, both innate and acquired, in feedback loops that go in different causal directions between host and microbiota (Hooper et al. 2012; see Pradeu 2012 for a philosophical treatment of immunology that includes microbes).

What is of considerable scientific and medical interest in such interventions is the resilience of the native microbiota and their ability to rebound after antibiotic perturbation and diversity decimation. This reversion is to a functional state similar to the one that existed prior to the intervention, even if some taxa remain depleted (Manichanh et al. 2010). Another remarkable finding is that microbiota disturbance can have neurochemical, brain developmental and behavioural effects on mice, as well as correlations with human neurological states (Cryan and Dinan 2012). These effects extend into all manner of evolved animal behaviours – from sexual preferences to predatory strategies (Ezenwa et al. 2012). Giant pandas, for example, appear to eat what their gut bacteria dictate they should: bamboo – which is high in cellulose, for which the panda's genome has no digestive enzymes – rather than what they are metabolically able to, which is other animals (Zhu et al. 2011). A range of mammalian and other vertebrate behaviour, including kin recognition, community and mate choice, is thought to be influenced by resident microbes (Archie and Theis 2011).

Microbial dynamics in these relationships can be greatly influenced by phage activity (Metcalf and Bordenstein 2012). In many animal lineages, phylogenies of gut microbiota branch with phylogenies of the host, thus indicating evolutionarily persistent relationships and the reconstitution of the community in each animal generation (Ochman et al. 2010).

Microbiomes have been made particularly famous not just by the thousands of human microbiome studies in the scientific literature but by how they have been taken up with enormous interest by a wide range of professional media. There are several reasons for this attention: first, the recognition of microbiota has challenged standard conceptions of what a human body is (due to the greater number of cells in a human body that are microbial than are strictly human); second, the concept of 'bacteria' (as a generic name for anything prokaryotic – see Chapter 2) is transformed in this recognition, because it requires that microbes are understood not just as harmful germs but as something helpful and even intrinsic to human health; and third, because microbial recognition appeals to conceptions of harmonious biology and connections to the rest of life that are not often prominent in traditional medical and microbiological literature but which have a long tradition in folk medicine and some older scientific accounts.

But are these microbiota the community, or does the community consist of the host and the microbiota as one unit? Both conceptions of community are possible. Metagenomics examines the microbiota via the microbiome and in the process learns a lot about the functional and dysfunctional interactions of microbes in environments such as the human gut. These interactions are embedded within further interactions between human cells and signalling molecules that have selective effects on the microbial cells and the human body (Schluter and Foster 2012). Both dynamics of interaction are reliably reproduced by different mechanisms over generations not only of the microbiota but the host-microbe unit as well (Fraune and Bosch 2010; see Chapter 4 and below). I discussed the classification issues of these interactions in Chapter 3, and the obesity, malnutrition, inflammatory disorders, diabetes and numerous other syndromes that have been interpreted as effects and sometimes causes of microbiota interactions. In addition to these bidirectional physiological interactions, there are high rates of LGT between gut microorganisms, and this rate is driven by the ecological roles these organisms fulfil, rather than phylogeny (Smillie et al. 2011).[9]

---

[9] However, see Skippington and Ragan (2012) for particular organisms in which phylogeny determines gene exchange.

Community-level selection is often invoked to explain the maintenance of functional effects despite taxa variation (e.g., Dethlefsen and Relman 2011. But how cohesive a community is the one in the human gut? Does it really work as something with a division of labour, an entity that has functional specialization and is reproduced over generations and across environments, with community-level properties selected *for*? This certainly seems to be indicated or is at least assumed by metagenomic research, but for most evolutionary theorists, the heritability of community properties would need to be mechanistically modelled and experimentally confirmed. I will discuss reproduction and inheritance further below, but for now, the short answer is that because functions are reliably reproduced by particular communities, over the short and long run reproduction of the community and inheritance of its functional properties is taken to be demonstrated.

Function and its persistence are, therefore, crucial to understanding the sort of assemblage a microbial community is. Mostly, in microbial ecology this has to be done on the basis of genes, with gene-regulatory, protein-interaction and metabolic networks reconstructed and biochemical functions inferred on the basis of those genes (e.g., Borenstein 2012). But unless the whole functional system is very simplified (that is, experimentally, or in a metabolically restrictive situation such as acid mine drainage), thousands of genes may not be sufficient to capture the biochemical processes in a community of interactive metabolizers (Ward 2005). As in all biological systems, there appears to be enormous redundancy whichever way microbiomes are analysed. Very different community compositions are thought to 'converge' on approximately the same functional states (Hamady and Knight 2009; however, cf. Allison and Martiny 2008). For many metagenomicists, the aim is to find the core microbiome or core community genome, which is the one shared by all the consortia in a defined habitat. This notion of a core, along with the enterotype idea mentioned in Chapter 3, are strategies used to decomplexify the diversity inherent in metagenomic datasets.

For some analyses, core identification is done on the basis of OTUs (e.g., Shade and Handelsman 2012), whereas for others, the core genes of the whole community are the focus (Turnbaugh et al. 2009; Burke et al. 2011). In the latter case, the type of biodiversity that is crucial to understand is not of taxa but of functions. Groups of organisms may be functionally equivalent, and as long as the functional niche is occupied, the taxa can be random arrivals (Burke et al. 2011; see below). Redundancy, however, may apply whichever way a core is conceptualized. Certain taxa and genes are often over-represented in OTU counts, implying functional redundancy (Shade and Handelsman 2012). But with studies showing that function can be imputed to community-scale genetic networks, not particular lineages of organisms, the attention therefore shifts to interacting genes rather than taxa. However, when function, and particularly energetic function, is the focus, in traditional plant and animal ecology this sort of research has come under the banner of ecosystem research.

Ecosystems are more inclusive than communities are because they include the whole environment in which communities function. For example, soil ecosystems include the chemical and physical constituents of the soil in addition to the microbes and other soil-dwelling organisms (Crawford et al. 2005). As already mentioned, Tansley (1935) coined the word 'ecosystem' as he tried to find a compromise in the debate about the status of communities. He thought the concept of ecosystem and its inclusion of 'inorganic factors' was required to understand the 'relatively stable dynamic equilibria' of such entities (Tansley 1935: 306). Ecologist Eugene Odum (1913–2002) made the notion of ecosystem even more dynamic than its earlier conceptualizations by focusing on the energy flows between trophic levels via mathematical models and equilibrium frameworks. Microorganismal processes were prominent in these models, and microbial ecology was identified as a contributing field (Odum 1971). Although the 'systemness' of ecosystems – as stable self-regulating entities – suffered the same fate as 'community' and fell into disfavour more than two decades ago, the use of the word and much of the older modelling machinery associated with it have endured.

The ecosystem notion has considerable importance as an organizing concept in microbial ecology even when the field is considered separately from mainstream ecology. In the textbooks *The Microbe's Contribution to Biology* (Kluyver and van Niel 1956) and *The Microbial World* (Stanier et al. 1957), the Delft School approach to microbiology encompassed not just microbial ecology for microbiologists but the reasons why biologists too had to care about microbial ecology. These reasons pivoted around the roles microorganisms played in shaping environments of every sort, but also took into consideration the way in which environments influenced communities. Microbes, hosts and environments need to be understood together for a full-fledged microbial ecology, urged early manifesto statements of the field (e.g., Steinhaus 1960). Today's microbial ecology may be in the process of bringing closer together traditional ecosystem (biogeochemical) and community (organismal) approaches than plant and animal research has so far managed (Loreau 2010).

Biogeochemistry is ecosystem analysis at its most generalized. It models biologically generated chemical fluxes in particular environments with the aim of understanding their interactions via organismal 'compartments' (Odenbaugh 2006). This is one area of classical ecological research in which microbes have been visible and integrated into modelling practice (another, as Chapter 6 will discuss, is ecological experimentation). Their visibility occurs because even though biogeochemistry mostly conceives of organisms as anonymous nodes of energy transfer and nutrient production, microorganisms have for some decades now been known as essential conduits of planet-wide processes such as the synthesis of organic matter, nutrient turnover, chemical transformations and climate effects (Ducklow 2008; Doney et al. 2004). Moreover, most biogeochemical transformations depend on multitaxon consortia of microorganisms. Not only do these collectives switch metabolic gears in different environments; they also exchange genes for valuable metabolic capacities. The interoperability of such genetic resources on the one hand enables individual lineages to flourish in different environments and on the other, communities to be compositionally flexible if function is not tied exclusively to particular lineages. For these reasons, there are now many calls to integrate biogeo-chemical modelling with metagenomic data about microbial activities (e.g., Reed et al. 2014).

Ecoystem perspectives can greatly expand how microbial communities are understood:

> The combined activities of microbial communities shape the face of the biosphere on a global scale. The power of these communities lies hidden in the metabolic versatility of their component species that, acting together, regulate ... the vast majority of matter and energy transformations on Earth. In a loose analogy, the entire biosphere can be imagined as a sort of 'superorganism' (Committee on Metagenomics 2007: 19).

Wholesale subscription to Gaia concepts is not what is going on in microbial ecology today, but two things are common to the ways in which 'community' or 'ecological system' are used. First, mixed-species microbial assemblages are often recognized as more than mere aggregations of individuals. As discussed above in the notion of community, there are causal properties that these organized systems possess, and microbial ecologists want to understand their effects, how the communities or ecosystems came to have these causal powers and how they persist. Second, acceptance of these larger collective entities does not come pre-loaded with anti-Darwinian thinking as it has in earlier centuries (see Sapp 2004).

In fact, just the opposite obtains. Although only a small amount of work so far has been done on how these multiscale entities evolve, it is taken for granted that Darwinian processes in particular and evolutionary ones in general apply to communities and ecosystems, even though there are still many doubts about selection at those levels (see Chapter 4). These doubts exist partly because of difficulties in using the standard conceptual machinery of organism, multicellularity and reproduction in an account of community-level selection.

Redefining organism, multicellularity and reproduction

The ubiquity of symbiosis and cooperation, and the communality of genetic resources – both revealed now even more clearly by molecular approaches – affect the way in which researchers think about biology's most taken-for-granted entity, the organism. Microbial ecology shows how collaborative interactions at many scales blur the usual distinctions that are made between so-called individual organisms and the larger organismal groupings of which they are part (see Chapter 4). Traditionally conceived organisms are biological systems possessing unique genomes, but many of the examples above and in Chapter 4 indicate that conceptualizing single-lineage organisms as autonomous individuals is misleading. Instead, the community and ecosystem perspective of microbial ecology suggests that the basis of any attribution of autonomy has to be function, which in many system-scale processes is the product of collaborative interaction. From this perspective, the boundaries between one traditionally conceived organism and another appear to be vague and defined by research interests (not meaning these aren't concerned with real things, however). If it is conceivable that the regulated metabolic collectives interacting with environments are actually 'metaorganisms' to some degree, then the next step might be to wonder about which sort of organismality is more fundamental and which is more derived. In other words, the collective might be a more basic state than the traditionally conceived isolated individual. This inversion of standard thinking draws useful attention to the broader view of biology that microbial ecology is emphasizing.

But if anyone is willing to take this step of recognizing more extensive 'organisms', a microbial ecological perspective also has implications for how multicellularity is conceptualized. Although most people – from both biologically trained and untrained perspectives – might think that multicellularity is an empirically obvious category, confidence in the notion of a clear line between unicellularity and multicellularity[10] is likely to dissolve when an ecological perspective, particularly a microbial one, is brought to bear on life. As I discussed in Chapter 4, all traditionally conceived multicellular organisms function with the inherited assistance of endosymbiotic partners, and these life-defining partnerships exist in interplay with numerous other forms of interaction (for example, plant-fungi-bacterial symbioses). Even single cells are complexes of interacting entities with considerable genetic autonomy, including prions, plastids, organelles and viruses (Dupré and O'Malley 2009). Although I speculated in Chapter 4 that Leibniz's notion of 'nested individuals' (Nachtony et al. 2002) might be helpful to understand these relationships, in many ways an ecological perspective undermines this Russian doll metaphor. The entities involved are highly interactive and they participate in a variety of multiscale activities. Stacking them one within the other and forming a traditional hierarchy becomes rather unsatisfactory in this light.

Because this interactivity is central to aliveness, most unicellular organisms are found, therefore, in collective lifestyles, several of which were discussed in Chapter 4. There are numerous examples of multicellular 'unicellular' organisms of the same species (for example, filaments and colonies). They exhibit coordinated growth, reproduction, sensing and movement (Shapiro 1998; Aguilar et al. 2007). Biofilms, metabolic partnerships and symbioses of multiple sorts are complex, highly structured aggregations that interact in regulated ways to metabolize, develop, reproduce and defend themselves. Many of these multispecies collaborations involve large organisms, which interact in multiple life-sustaining ways with resident microorganisms. If the concept of multicellularity can be broadened to include multilineage communities, where do viruses fit in this schema? A very important boundary for most discussions of life, whether from ecological or evolutionary perspectives, is the one between cellularity and non-cellularity. As I mentioned in Chapter 2, life is usually defined as something that is exclusively cellular, but microbial ecological insights suggest that this is not the only or the most appropriate criterion.

---

[10] In addition to unicellularity and multicellularity, another form of cellular organization needs consideration: syncytia and coenocytes (Stanier and van Niel 1962). Syncytia are extended cells due to processes that remove the cell walls between groups of individual cells; coenocytes are cells in which full cell division does not occur but more nuclei are produced. These cell types are common across the eukaryotic world, and found in plants, fungi, algae and protists as well as animals (in some special situations or bodily locations). Syncytia and coenocytes have multiple nuclei, sometimes of different genetic composition (for example, in the arbuscular mycorrhizal fungi on plant roots, which were discussed in Chapter 4). These types of cell are sometimes suggested to be important to evolutionary transitions in individuality (e.g., Grosberg and Strathmann 1998).

Viruses are implicated in every life form and process; they are indirect beneficiaries of all metabolic processes; they play major evolutionary and ecological roles throughout the history of life. Metagenomic and other sequence data have given remarkable insights into the extent of the diversity, ecological function and evolutionary role of viruses in the cellular world. Whether viruses are community members or ecosystem components (that is, abiotic), they, too are participants in these larger communal organisms.

As mentioned already in the discussion of communities, reproduction is usually the obstacle that is thought to block multilineal communities from being conceptualized as units of selection. This is because of how reproduction is considered crucial to the Darwinian evolution of organisms, in that it is responsible for transmitting the inherited properties essential to fitness (Godfrey-Smith 2013). From a strict Darwinian perspective, multispecies arrangements of cells may be doing many things, but because they do not reproduce in any unified bottleneck-producing way, they fall foul of any reasonable definition of organismality. As neo-Darwinian Richard Dawkins claims, 'The essential, defining feature of an organism *is* that it is a unit that begins and ends with a single-celled bottleneck. If life cycles become bottlenecked, living material seems bound to become boxed into discrete, unitary organisms' (Dawkins 2006 [1976]: 264).

Bottlenecks allow a 'fresh start' because of reversion to the germline and loss of somatic mutations and many parasites; bottlenecks also permit small mutational changes in the germline to have major effects on the organism's descendants. Dawkins says all this from the point of view of a zoologist. Imagine he is right. Although single lineages of microbes might be considered to have acceptable bottlenecks (they are unicells, after all), the collectives I have been discussing definitely don't.

Philosopher Frédéric Bouchard proposes one solution to this impasse. Differential reproductive success, he says (2008), simply does not capture the evolutionary fitness of some forms of biological organization such as colonial organisms, symbiotic communities and ecosystems. The persistence of such entities has to be understood as indicative of fitness, rather than progeny counts of individual (traditionally conceived) organisms. Many collectives of organisms, or ecosystems when abiotic factors are included, persist over lengthy periods of organismal time, and thus would not need to demonstrate bottleneck-like reproduction to prove they were 'organisms' in the Darwinian sense. Moreover, community-wide activities are crucial conditions for the reproduction of any component. Many multilineage consortia arise anew, persist and regenerate in the process of producing collective metabolic advantages. But because these varieties of multicellular entities do in fact give rise to new generations of similarly functioning biological entities, persistence is probably only part of the answer.

'Collaborative reproduction' might be a second answer to objections to designations of organismality on the ground of reproduction. Multilineage organisms do indeed reproduce, but not necessarily via a bottleneck or simultaneously, except in the case of obligate prokaryote symbionts (Moran et al. 2008). However, they do make more of themselves to perpetuate the relevant functional properties of the larger collective. Note that Dawkins says, '*if* life cycles become bottlenecked' (emphasis added), then standardly conceived 'discrete, unitary' organisms are the result. In the cases of communities, bottlenecking does not occur, so another type of organism is arguably being reproduced. Reproduction in such collaborations may involve both vertical and horizontal strategies, via sexual and asexual mechanisms, such as the combinations employed in the symbioses that create lichens (Grande et al. 2012). Even these key-partner symbioses – such as lichens formed by fungi and algae or cyanobacteria – are also partnered by much more diverse prokaryote communities. These 'extra' community participants are just as intrinsic to the focal cooperative entity and are persistently reproduced by processes additional to those employed by the standardly identified organisms (Hodkinson et al. 2011). Moreover, the heritability of community phenotypes is a topic that is being examined with some success, both theoretically and experimentally in microorganismal communities and plant ecosystems (e.g., Swenson et al. 2000; Peiffer et al. 2013). And of course, both reproduction and functional inheritance can fail in any multilineage system, just as they could for the entity-based regeneration of any unilineal metazoan, for example.

In a recent treatment of reproduction in relation to organismality and Darwinian individuals (also known as units of selection – see Chapter 4), Godfrey-Smith (2013; 2009) outlines three dimensions of 'collective reproduction'.[11] He is referring to single species, however, and not the 'collaborative' reproduction I mentioned earlier. These dimensions vary along axes of bottlenecks, reproductively specific cells and overall integration. Biological entities can score differently along these axes and thus be more or less marginal organisms and Darwinian individuals. The example of collective

---

[11] Other types of reproduction in Godfrey-Smith's (2013) schema are 'simple', which covers most unicellular organisms considered in isolation, and 'scaffolded', which is how viruses, genes and mobile genetic elements are replicated.

reproducers that scores perfectly is humans (they are collective because they are groups of cells) but this scale of collectivism does not include the microorganismal complement to human cells. Godfrey-Smith's three dimensions provide a very useful scheme for understanding some central biological and evolutionary processes, but it seems unlikely that the best starting place for an analysis of reproduction would be humans understood in isolation from microbes. This is because there is so much evidence *against* collections of animal cells and animal cells alone as the most basic, important or even most biologically common entity.

A perspective informed by microbial ecology would be much more likely to start with the multilineage collective as the exemplar, and to score other arrangements in relation to how they depart from that exemplar. Humans and many other animals would be very interesting special cases in that schema, but would not be the template against which all the rest of the interacting biological world (of which there is far more and far older forms) should be compared. Criteria such as the collaborative capacity to produce another generation, functional persistence despite variability in components, and the maintenance of life-sustaining causal connections are likely to be central to such scoring. But even if we were to agree to start from humans or other animals, Godfrey-Smith's schema makes it clear that nothing in biology says reproduction can happen in only one mode. Instead, it says that detecting organismality is a matter of working out for any system – including the human-microbiota system – what modes of reproduction are at work. Godfrey-Smith's (2013) discussion of this topic then goes on to include multispecies units as organisms, which in very close collaborations are Darwinian individuals, too (for example, aphid-*Buchnera* systems).

A broad eco-microbiological view allows a compromise. Many standard units of taxonomy, evolution and ecology can be extracted from a microbial matrix and given context-specific conceptualizations and definitions. If a Dawkinsian sort of organism is required, microbes can be demoted to the 'environment'. However, if a more realistic view is wanted of how living things work in ecosystems, a multiscale microbial perspective is the better starting place. And if general accounts of organism, individual, multicellularity and reproduction are the aim, then philosophers will do better by taking a microbes-first perspective before focusing on their favourite animals.

Theory and modelling in microbial ecology

In light of all the technological innovations described above, it might seem straightforward to say that microbial ecology has been methodologically and technologically driven, from the elective culturing regimes Winogradsky and Beijerinck devised, through the innovative sampling and in situ culturing devices made by ZoBell and Jannasch, to the molecular tools pioneered for environmental DNA by Pace and others. However, although microbial ecology is an illuminating example of a field in which technology innovation is inseparably interwoven with the production of knowledge, it is also woven through and through with theoretical concerns and developments. Researchers such as Hungate were immensely concerned to quantify organisms and their biochemical reactions in natural environments in order to build more accurate mathematical models (Hungate 1966). Biogeochemistry is largely concerned with tracking flows of energy and nutrients and does this with a variety of models, most of which are mathematical or computational (Doney 1999). Brock had conceived of hot springs as mathematicizable steady-state systems (Brock 1995) and he argued repeatedly that progress in microbial ecology would depend on 'the development of quantitative models that will not only describe the rates of growth and activity of ... microorganisms, but which will also describe these processes under a wide range of environmental conditions ... We *must* develop general principles that will be widely applicable' (Brock 1987: 13).

But first, it is helpful to establish why theory might be necessary in microbial ecology. Are current demands for it (e.g., Prosser et al. 2007) blindly imitating classical ecology, which has a long history of preoccupation with mathematical modelling? This seems doubtful, whichever way we look at microbial ecology. Many microbial ecologists believe that the very nature and scale of microbial systems mean that the descriptive tools currently dominating the field are insufficient for real progress. Technology, likewise – it might demonstrate ingenuity in its application, but ultimately needs to produce more than descriptions of components such as genes, organisms and metabolic pathways (Curtis 2007). Genuine understanding of what is going on microbially in any ecosystem will require 'a calibrated set of rules to describe and predict the behaviour of the microbial world as a system' – an achievement that may well go further than standard ecology with its experimental constraints and 'mathematical castles in the air' (Curtis 2007: 1; Prosser et al. 2007). In recent molecular microbial ecology, multiple research projects do in fact take the necessity of theory for granted and aim to model their ecosystems based on quantitative data. Some theoretical efforts extract broad conceptual threads, such as the nature of ecological units (as discussed above), or provide general explanations of evolutionary and ecological processes in microbial ecosystems, but many others build on and extend traditional models already tested and accepted (or sometimes still contested) in plant and animal ecology. Although there is a range of models applied from macroorganismal ecology to microorganismal,[12] I will focus on one general area of application to show the conceptual tensions and methodological innovations that arise in these efforts.

Species-area relationships, biogeographical distributions and community assembly rules

Biodiversity estimates are the hard currency of plant and animal ecology. But organism or species counts are carried out in order to detect patterns that can then be explained or modelled theoretically. Classic ecology has paid a considerable amount of attention to how organisms are distributed at different spatial scales. In particular, the relationship between species and area was noted early in natural history – from the late eighteenth century onwards (McGuinness 1984). Scatter plots of this pattern depict a linear relationship between species and area: when area increases, so does the number of species. The relationship was mathematicized in 1921 as a quantitative power model with a fitted constant (see Lomolino 2000a for a discussion of this model's problems). Why is this relationship important? For animals and plants the species-area relationship captures a major pattern that allows predictions about biodiversity to be made, and thus calls for explanation of those patterns. It is often cited as one of biology's few laws or universal rules (e.g., Lomolino 2000a; Horner-Devine et al. 2004), but it is still not clear whether and to what extent it would apply to microbial diversity. In microbial ecology, one of the main reasons to try to apply the species-area relationship to microbial communities is to see whether microbial patterns are the same as macrobial ones, and then to find out what the explanations for the similarities and differences are.

The species-area relationship is a monotonic relationship that can be plotted as a slope, called 'z' (Figure 5.2).[13] Plant and animal species-area relationships are fairly steep in slope because their diversity differentiates steadily with area. Although widely believed to hold in any plant or animal

---

[12] For example, succession theory – the dynamics and structure of colonization of an environment – (e.g., Jackson 2003), diversity-productivity relationships (e.g., Smith 2007) and the fundamental niche concept (e.g., Materna et al. 2012) have all been applied from macroorganismal ecology to microorganismal ecology.

[13] The species-area relationship is often called a taxa-area relationship, because it can be applied at different levels of the taxonomic hierarchy; also, when species are tricky then taxon is a more ambiguous substitute.

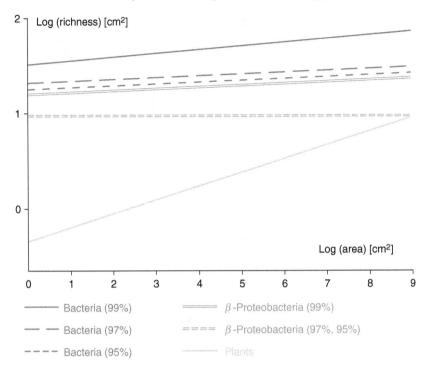

Figure 5.2: Species-area relationships and the *z*-slope, comparing macrobial and microbial plots. The standard steep curve of plants is very much in contrast to the flatter slopes of bacteria, whether in general or of particular taxa. Percentages refer to the level of sequence similarity used to designate OTUs. Based on Horner-Devine et al. (2004).

distribution, many communities do not in fact fully realize the distribution patterns captured by this slope (Lomolino 2000b). In microbial ecology, a number of studies also do not find exactly the same or even any steepness of slope as a function of richness of species and size of area (e.g., Green et al. 2004; Horner-Devine et al. 2004; Figure 5.2). Other studies, however, do (e.g., Bell et al. 2005; van der Gast et al. 2006). Early findings that prokaryote diversity did not greatly increase with area were explained away as an artefact of poor taxonomic resolution (species having been defined too grossly so less diversity was found), habitat specificity (many prokaryotes managing to survive in a wide range of conditions so starting off more broadly distributed anyway) and scale (microorganisms needing to be studied at very tiny and very large spatial scales, but not the scales that fit plants and animals) (Horner-Devine et al. 2004).

Although even the weaker findings of a flatter curve in prokaryotic and eukaryotic microorganisms are interpreted to support the existence of this 'universal law' (Horner-Devine et al. 2004: 752; Green et al. 2004; Reche et al. 2005), there are reasons to doubt that current data are sufficient to establish this universality. All the sampling and diversity calculation difficulties mentioned earlier affect the study of species-area relationships (Prosser et al. 2007; Woodcock et al. 2006). The notion of the rare biosphere, in which low-abundance organisms are simply not sampled, means that diversity estimates for any area would not reflect the existence or distribution of those organisms. The slope could in fact be steeper if these rare organisms were detected. Another factor affecting how the relationship slopes is the ability of many microorganisms to survive in dormant phases. Dormancy affects microbial dispersal and immigration by increasing it over larger areas, but then decreases organismal diversity as area is increased (Lennon and Jones 2011).

If there really were little diversity increase over geographic distance in microorganismal distributions, this would appear to mean that geography plays only a minor role in explaining distributional patterns in microbes. Biogeographical concerns have a well-known history in microbiology because of the fact that historically, the distribution of microorganisms across the Earth was thought to be unrestricted by geography. In Baas Becking's words, built on Beijerinck's, '*Everything is everywhere*, but, *the environment selects*' (Baas Becking 1934; Beijerinck 1913; O'Malley 2007). The small size of microorganisms was meant to enable their limitless dispersability ('cosmpolitanism'), and only the need for particular environmental conditions would mean they were found at some locations and not others ('endemism'). In other words, geography would have little or nothing to do with either the distribution patterns of microorganisms or their speciation (see Chapter 3). Quite a few molecular studies confirm this assumption. As these studies test the species-area relationship, they find far less increase in microbial biodiversity over geographical distance than there is of plants and animals (e.g., Fierer and Jackson 2006; Figure 5.2). Environmental variables, such as acidity or alkalinity, or seasonal changes, make the difference (Gilbert et al. 2012).

A fair amount of other molecular ecological data is, however, telling a different story and revising the 'everything is everywhere' assumption in at least some circumstances. A number of microbial groups, prokaryotic and eukaryotic, are now known to have endemic distributions – confined to particular geographical locations – even in highly similar environments (Papke et al. 2003; Whitaker et al. 2003; Martiny et al. 2006). Different ocean depths, for example, may show very different distribution patterns with surface communities shaped by environmental conditions and deeper communities structured more by geographic factors (Ghiglione et al. 2012). The cosmopolitanism position is still maintained, however, because many of the differences in population distributions are found at subtle genetic levels and are sometimes argued to be phenotypically irrelevant (e.g., Finlay et al. 2006). But as already discussed, the rRNA molecules often used in such studies, including those of protists, are too coarse to capture important ecological differences between organisms, and relying on them diminishes diversity and obscures biogeographical patterns (see Chapter 3). And when metagenomic analyses examine the biogeography of functional genes with no reference to the gene bearers, they find that 'all *functions* are everywhere, but the environment selects' (Raes et al. 2011: 4).

The debates here turn on what matters most for deciding whether sequence similarities and differences are the right proxies for the ecological diversity of organisms, and what the mechanisms are (selection, drift, dispersal, mutation) that produce biogeographical patterns, whether of ubiquitous or restricted distribution (Hanson et al. 2012). The assumption that anything microscopically small must have unlimited capacity for dispersal has been challenged: dispersal requires a multifactorial approach especially in microorganisms (Jenkins et al. 2007). Even within protists there are different dispersal capacities due to different cell sizes (Wilkinson et al. 2012). Alternative explanations may be required for prokaryotes as opposed to microbial eukaryotes, and for different functional groups within prokaryotes (Ragon et al. 2012; Lindström and Langenheder 2011). And even if functional genes are the focus, it is overly adaptationist to presume that everything needed by organisms in a particular environment will turn up (that is, all functions are everywhere), whether by mutation, LGT or endosymbiosis (see Chapter 4). Demands for more precise quantitative measures and experimental studies to establish mechanisms continue to be central to the development of detecting and explaining patterns in microbial biogeography (Martiny et al. 2006; Lindström and Langenheder 2011).

The species-area relationship goes beyond simple demonstrations of slope and the questions it raises about whether geography matters to these patterns. The relationship is the 'stepping stone' to subsequent theories or explanatory accounts of how communities assemble in any area. In plant and animal ecology, neutral theories of community assembly vie against niche-based theories of assembly. In the latter, it is assumed that species compete to occupy niches to which they are adapted. In neutral theories, however, dynamic equilibrium between immigration and extinction is the force that brings about community composition. The early version of this account is the famous 'island theory of biogeography' that Robert MacArthur (1930–1972) and E. O. Wilson formulated in 1963 and elaborated in 1967, with the help of (amongst other data) a little evidence from microbial experiments (MacArthur and Wilson 1967: 42–3). 'Where microorganisms are concerned', they said, 'the "islands" can be created artificially' (MacArthur and Wilson 1967: 182). This is a claim I will take up in Chapter 6 when I discuss how microbes can be used as tools and models for macroorganismal research. In the initial account of island biogeography, the number of species does not change, although the species composition of the community may. The balance is a product of how immigration and extinction work across the community. More isolation means less immigration, but bigger areas mean less extinction. This model of community composition has major implications for conservation theory because it suggests that it is better to preserve fewer larger areas rather than many smaller ones. The problem is that there are limited or only contentious data to support equilibrium theory in this form, and even former advocates of island biogeography eventually questioned its empirical and theoretical adequacy (Simberloff 1976; Lomolino 2000b).

As a response to these problems, forest ecologist Stephen Hubbell devised in 2001 a 'unified neutral theory of biodiversity and biogeography'. It emphasizes the functional equivalence of trophically similar species and, again, explicitly refuses the relevance of niches to the structure of communities. Instead, birth, death and dispersal are the key factors to be considered, and interspecies dynamics are a zero-sum game (Hubbell 2001). In other words, no species can increase without a matching decrease in the biomass or abundance of other species. There have been many criticisms of this model, mostly on empirical grounds (e.g., McGill et al. 2006). Its update steps back a bit to urge ecologists to see the neutral theory of island biogeography as a 'null model', in which finding it not to be true is still very useful. In addition, the notion of ecologically equivalent species is demoted to a 'simplifying assumption' that is not required to be strictly accurate (Rosindell et al. 2012: 203). Perhaps even more significantly, however, this general model was developed solely in consideration of plants or animals at single trophic levels, almost always having small populations, extensive sampling and known species-abundance relationships (Caruso et al. 2011; Curtis et al. 2006). Seeing whether it applies to the assembly of microbial communities – which occupy multiple trophic nodes, have large populations, poor sampling and totally unknown species-abundance relationships – has become a very active research area in microbial ecology.

Islands are easy to find or even make for microbes, as MacArthur and Wilson noted (1967). Microbial ecologists have thus been inspired to study some microbial habitats as islands. For example,

> We suggest that tree holes and similar habitat patches are islands of relative stability where microbial communities can approach equilibrium. Under such conditions, the patterns of abundance and diversity of microbial communities would be similar to those found for larger organisms (Bell et al. 2005: 1884).

These tree hole and other 'island' studies have provided strong support for both species-area relationships and neutral accounts of that pattern (Sloan et al. 2007) – stronger support than found for macroorganisms, in fact (Woodcock et al. 2007). But these findings have been challenged too, and an important case to bear in mind is that human bodies, especially the intestines, do not seem to be islands populated by communities that are assembled neutrally (Walter and Ley 2011). At least some of the differences are due to the many host interactions that shape community composition and function in these habitats. Biofilms in environments such as streams also do not show patterns of stochastic assembly but of species sorting, which is niche dependent (Besemer et al. 2012). In addition, the focus on species distribution rather than functional distribution is contested: examining the spatial distribution of functional traits rather than taxa is likely to be more explanatory (Green et al. 2008).

Meta-analyses of different microbial biogeography studies indicate that different mechanisms of community assembly might work for different groups in the community, whether functional or taxonomic groups (Lindström and Langenheder 2011; Lankau et al. 2012). Other studies and reviews have also detected, even in fairly simple systems, considerably varied combinations of neutral and deterministic mechanisms of community assembly in relation to different spatial scales and functional groups of organisms (Caruso et al. 2011; Fierer et al. 2012b; Stegen et al. 2012). This is unsurprising to the researchers involved because focusing on a single trophic level, as the original neutral models do, is not very realistic. Moreover, scale is again crucial, and diversity calculations and models have to take into account a much greater range of spatial units in microbes than in most plant and animal ecology (Grundmann 2004). The ability to range across different nodes of nutrient supply and consumption, and to interact at different scales, is a capacity that has to be factored in when modelling the role of microbes, whether for the purposes of explaining biogeochemical cycles or assembly patterns. Metacommunity theory, which distinguishes between dispersal, local diversification, environmental selection and evolutionary drift as forces that structure communities, offers not only theoretical insight into the multiple processes in microbial community assembly but also provides guidance for clinical practice (Costello et al. 2012; see the discussion of disease ecology in the conclusion to this chapter).

But if we were to agree that results showing at least some neutral assembly processes are reliable and realistic, we then need to come to grips with how consistent these results are with the idea of communities or even ecosystems as causally efficacious, cohesively organized systems (see above). A very intriguing molecular study of the bacterial communities associated with tidal-pool algae has managed to synthesize both niche and neutralist perspectives. It proposes the existence of a niche that is colonized by organisms able to carry out particular functions, but because these organisms are functionally equivalent in regard to that niche, their colonization is random: first come, first served (Burke et al. 2011; van Ooij 2011). The phylogeny of these organisms varies considerably across sites, but a core set of metabolically appropriate genes is consistently found despite the organismal variation. Because it takes an analysis of genes to capture this dynamic, Burke et al. (2011) suggest that this may be a key difference between microbial and macrobial ecology – with the latter often being understood solely on the basis of species counts. However, given there is so little eukaryotic metagenomics (see the earlier footnote), it is not yet possible to compare similar datasets. Moreover, to assess the proposed explanations properly would require multiscale analyses of interactions – simply categorizing community structure as neutral or niche-based (or most likely, both) is only the beginning.

But the differences in macroorganismal and microorganismal ecology to which this study points are worth considering further. Many well-established tools and methods in macroorganismal ecology are being put to use in microbial ecology, sometimes with further development of those tools. Despite many success stories, it is still not clear that what works for large-organism ecology should work for microorganismal ecology. As noted, the scales of biological activity are immensely different (Konopka 2006; Grundmann 2004), and the interactions between organisms occur in very fine-grained and intimate relationships. Microorganisms are not restricted to one scale of activity and their impact is more likely to be at a variety of scales of organization – from micro-environmental molecular transfers to biosphere-level biogeochemical cycles. Macroorganismal ecology is carried out with species as the main units simply because this is the most obvious way to proceed. By taking out of necessity the molecular route, microbial ecology may in the end be able to teach macrobial ecology new tricks, just as microbial phylogeny did to macrobial phylogeny. This was not just about the detection of LGT, but the use of molecular methods in general. A major clarification of animal phylogeny, for example, was achieved on the basis of microbial phylogeneticists collaborating with animal phylogeneticists whose previous focus had been morphological (Field et al. 1988).

Another difference between macrobial and microbial ecology might be that microbes themselves are explanatory, in the sense of explaining large-organism abundances and relationships. Even if the ecological equivalence of species is accepted (as neutral theories require), the force that structures any macroorganismal population might be microbial – not just pathogens but beneficial microbes as well (Ricklefs 2012). The plant-fungi-bacteria systems discussed in Chapter 4 are good candidates for such an explanation. However, it is very unlikely that host factors are not causal, and thus an adequately microbial explanation of community assembly will have to include multiscale interactions. Despite the many simplifications involved, it is still valuable for microbial ecology to establish its similarities to and differences from macrobial ecological processes. In the process, it is very probable that a more complex overall view of ecosystem interactions will develop, and microbial ecology will feed back into the models and concepts of macrobial ecology.

Concluding reflections on microbial ecology

Although often advanced as the antithesis of pure culture, today's microbial ecology brings together the different historical strands of its parent discipline. Van Niel's agenda of integration referred primarily to laboratory and field studies, but in contemporary microbial ecology, integration encompasses a wide range of approaches concerned with interactions between different scales of biological organization and the structure of biodiversity. For example, culturing can capture organisms not identified by molecular studies, including – paradoxically – the very rare ones that escape sequencing efforts, and metagenomic data can indicate appropriate culturing conditions (Shade et al. 2012; Singh et al. 2013).[14] One reason for microbial ecology's enthusiasm for community-oriented approaches is

---

[14] Schaechter (2012b) calls the success of culturing rare organisms 'paradoxical' because the very idea that extremely rare and previously undetected organisms would succumb to culturing approaches goes against many of the founding assumptions of metagenomics (see above).

because they foster an integrative methodology, and this has productive outcomes in terms of generating knowledge and leading to further research questions. With research groups focused on functional communities able to combine molecular, experimental and modelling methods all in the same study, major advances can be made in short periods of research, as the last decade of metagenomic-driven inquiry attests. The prospects for a truly general microbiology are better now than they have ever been, and may even in important respects encompass the biology of multicellular organisms. And despite microbial ecology's history of being seen as an alternative to medical microbiology, there is reason to think a rapprochement is not only possible but already happening.

In the next chapter I will explore the theme of how microbiology can either stand in for or include large-organism biology but for now, will simply emphasize the ways in which microbial ecology (including metagenomics) can work with medical microbiology. As the two perspectives on microbes come together, they realize a much older conception of 'disease ecology' (Anderson 2004), in which microorganisms are investigated via a natural history approach in order to understand their broader environmental roles in producing disease. The culture-independence of many recent molecular methods, as I suggested above, undermines a total reliance on Koch's postulates and single organisms as the causative agents of disease. Cystic fibrosis is an interesting example of a disease that microbial ecology is reconceptualizing. This lung disease is associated with four or five main taxa, but metagenomics and other molecular research have greatly augmented the number of organisms correlated with its disease patterns (Rogers et al. 2009). Immunity perturbations of the host are thought to be of major causal relevance, too (Sibley and Surette 2011). While taking a community orientation to disease can simply mean diagnosing a whole community rather than one taxon as pathogenic, this in itself already implies greater attention to interaction between microorganisms at the community scale and their host interactions.

Some of the human microbiome studies essentially follow this template in which 'pathogenicity' is associated with communities as causes of specific disease. However, coarse causal attributions ('it's the community') are not very helpful (see Relman 2013, and above), and more fine-grained studies have sought to explain syndromes such as obesity as the result of changing relationships in microbiota interactions with the host (see Chapter 3). Organisms that are minor in a compositional sense may be major from a functional-mechanistic perspective (Hajishengallis et al. 2012). And increasingly, contemporary microbial disease ecology seeks out more subtle and dynamic evolutionary and ecological patterns to generate mechanistic explanations and effective treatments of disease (Pepper and Rosenfeld 2012; Dethlefsen et al. 2007). Perhaps the ultimate payoff in new microbial disease ecology will lie in identifying and promoting 'healthy' communities – microbes in association with hosts – rather than targeting pathogenic ones (Costello et al. 2012; Dethlefsen et al. 2007).

Antibiotic interactions are just one example of a medically relevant phenomenon that is usefully understood ecologically rather than via simple host-pathogen conceptualizations. Thinking simplistically about wiping out particular pathogens is problematic because it does not take into account the evolution of antibiotics and their ecology. Antibiotic use is now a problem because of the increase in antibiotic resistance around the world (Davies and Davies 2010). Resistance genes can arise rapidly through mutation or the exchange of relevant DNA modules between organisms. The strong selective pressure of antibiotic treatments makes such resistance mutations and transfers beneficial to the organisms newly endowed with resistance. Even without current selective pressures, microbial communities harbour resistance reservoirs (Looft et al. 2012). Persister organisms (discussed in Chapter 4) can withstand antibiotic treatments for lengthy periods of time. But on top of the well-recognized problem of resistance is the one of microbiota perturbation and the fact that repeated antibiotic application – both deliberate and accidental, as when antibiotics get into wastewater or food items – can restructure a community. Despite the capacity of microbial communities to rebound from initial treatments, frequent antibiotic exposure can change community structure permanently, often with undesirable functional consequences for the host (Dethlefsen and Relman 2011). For example, mice whose gut microbiota are disturbed by antibiotics, especially penicillin, at developmentally sensitive times grow up with many metabolic differences from non-treated mice, and are much more likely to be obese (Cho et al. 2012).

If antibiotics are understood in terms of individual components alone – molecules or organisms, both large and small – the dynamics of resistance are poorly explained and predicted. But even understanding community-scale antibiotic dynamics does not fully capture how the production of antibiotics and antibiotic resistance occurs. For that, natural environments and evolutionary history need to be brought into the picture (Allen et al. 2010; Martinez 2009). Antibiotic resistance is known to be ancient, having quite possibly arisen three billion years prior to human overuse and release into the environment of antibiotics (D'Costa et al. 2011). Antibiotic molecules can be found in many natural environments that have no anthropogenic sources of antibiotics. Their function is usually understood from how they are studied in the clinic, under high-concentration damaging doses. It is not clear this sort of dosage is ever encountered in most microbial environments, where sub-lethal concentrations are much more the norm (Linares et al. 2006). Antibiotics have much wider effects than simply annihilating other bacteria in these lower concentrations. Some microorganisms use antibiotics as food (Dantas et al. 2008). Other probably more important functions of antibiotics are as signalling molecules. These signals could be manipulative of the host or work to benefit any nearby cells, whether host or microbial (Fajardo and Martínez 2008). Waksman, mentioned in the history of microbial ecology above, conceived of antibiotics as weapons of microbial war, but this account does not capture their evolutionary and environmental aspects (Davies and Davies 2010). The ways in which antibiotics and antibiotic resistance function will vary in dependence on the environmental context and the evolved relationships of the microorganisms with other organisms in those environments. To develop new antibiotics to deal with antibiotic resistance and new diseases will require more extensive community-based experiments and analyses (Martinez 2009).

This example of how a broader ecological perspective transforms a narrower medical perspective shows how microbial ecology is in the process of transforming organismal biology far beyond even van Niel's vision. This is not just epistemically, in how it allows knowledge to be generated about living things, but also ontologically, in how it allows us to understand what sorts of living things there are. Historians of microbiology have discussed the 'great golden era' of microbiology as either its early microbe-hunting days in the late nineteenth century or the first decade of molecular biology in the 1940s (see Chapter 6). This chapter's examination of molecular microbial ecology, however, shows how much further the reach of microbiology has extended into the life sciences and biomedicine. A number of microbial ecologists have argued for the immense value of microbial mixed systems for any form of ecological research (e.g., Jessup et al. 2004). Although tractability and simplicity are often the reasons given for using microbial model systems in ecology, a case can be made that there are also representational reasons for looking at microbial systems first and only afterwards extrapolating to large-organism systems. The discussion above of how to conceptualize organisms, multicellularity and reproduction should have shown that to a certain extent. In the next chapter, the last substantial one of the book, I will consider in several more spheres how microbiology has managed to become a vehicle for other non-microbial life sciences, and what the implications are for microbiology and other biological research.

CHAPTER 6

Microbes as model biological systems

This chapter investigates the way in which microbes have been used in general biological research. Microbial model systems have served as the material and epistemic bases for several incarnations of molecular biology, as well as experimental evolution and ecology. Microbes have been used to exemplify general theoretical accounts and concepts, pertinent to all life, and they do this via capacities for representation and tractability. Microbial model systems also illuminate philosophical analyses of classification, evolution and ecology. I suggest as a conclusion that philosophy should start with microbes as the entry point into biological reflection and only subsequently focus on larger organisms.

Microbes as model molecular systems

> Bacterial genetics has had more profound impact on the development of modern biology than any other biological discipline (Brock 1990: 1).

The idea that microorganisms could be used to understand any life form has important roots in the late-nineteenth century. Beijerinck – whom we met in the ecology and classification chapters – argued alongside founding geneticist William Bateson (1861–1926) that bacteria should be understood as genetic and biochemical model systems. He thought they were both experimentally tractable and representative of all forms of life (Beijerinck 1900–1901; Bateson 1907, in Summers 1991). Beijerinck's research focused on how different bacterial physiologies could be understood as adaptations to environmental conditions. He believed that microbiology could contribute to a universal theory of life that would be built on the new science of heritability and variation (1895, in Theunissen 1996; van Iterson et al. 1983 [1940]). This science eventually would be known as genetics, and Beijerinck predicted that enzymes would form the material basis of heritable 'genes'.

But for microbes to become model systems for genetics – regardless of what genes were thought to be – took longer than Beijerinck had imagined. Although microbiology was successfully installed in the first decades of the twentieth century as an academic discipline, using microbes to model living systems took a physiological focus first. In the emerging Delft School of microbiology in the 1920s and onwards, Beijerinck's successor, Kluyver, envisioned that the unity of life would be found in the biochemical processes of metabolism (1959 [1924]; Kluyver and Donker 1924). Kluyver saw the possibilities for a general microbiology that would integrate the applied findings of medical, industrial and agricultural microbiology, and bring 'order into this chaos of phenomena' (Kluyver and Donker 1925, in Friedman 2004: 53). He aimed to find general principles that united chemistry and biology. Microbes were the obvious starting point because they were the most physiologically diverse group of organisms. Kluyver announced this unified life science would show that 'from the elephant to butyric acid bacterium—it is all the same!' (Kluyver 1926, in Friedmann 2004: 58). Jacques Monod (1910–1976) and François Jacob (1920–2013) reworked this adage a few decades later as they mused over the implications of the genetic code. They speculated that 'anything found to be true of *E. coli* must also be true of Elephants' (Monod and Jacob 1961: 393).[1] This claim was a major motivational force in early molecular biology.

Major figures such as Delbrück and Joshua Lederberg (1925–2008), along with many other molecular biologists of the 1940s and 1950s, believed that molecular analyses carried out on microbial life forms would produce results generalizable to all life. Beginning with early observations in radiation genetics and guided by epistemic commitments to the unity of nature, Delbrück reasoned in the 1930s and 1940s that genes could best be understood as molecules. Even more radically, however, he was convinced that viruses, rather than *Drosophila*, were exemplary 'living organisms' because of their ability to replicate in a 'genes only' way, without the complexity of cell division (Delbrück 1969). This most basic reproductive process made viruses highly tractable model systems, he argued. Increasingly, other geneticists agreed, at least in regard to unicellular organisms. Geneticist Milislav Demerec (1895–1996), the scientific director of the Biological Laboratory of the Long Island Biological Association at Cold Spring Harbor (a historical hothouse of molecular biology) sought to reassure funders and administrators of the value in using microbes as general model systems.

---

[1] This was only a tentative claim, however, because Monod and Jacob were concerned that only certain aspects of the genetic code would be universal and others would be taxon and enzyme-specific.

You may properly ask . . . why we use microorganisms in our investigations. The answers to this are very simple. By using as experimental material these microorganisms, which are easy to handle, biologists and biochemists are trying to unravel two of the most intricate puzzles of all living matter; namely, the mechanism of heredity, and the reproduction of living substances. Since the fundamental laws of nature are general, discoveries made by working with these minute organisms help us to understand the life processes of higher living beings (Demerec 1946: 217).

Phage research produced a platform of research achievements that encouraged Delbrück and others to look more deeply into the complexities of gene replication and inheritance. In addition, research conventions were established around these organisms to help organize the nascent field of microbial genetics. Phage genetics was standardized by the choice of model phages, which reproduced themselves with the assistance of another rising star in the new world of molecular genetics, *E. coli*. This bacterium displaced the popular microbial fungus, *Neurospora*, as a source of major breakthroughs for understanding relationships between mutations and biochemistry (Beadle and Tatum 1941). Although it could be argued that microbial genetics was established when the role of genetic mutations in inherited microbial traits was confirmed (Luria and Delbrück 1943; Chapter 4),[2] methodologically, the hallmark of genetics had always been crossing experiments. This required the sexual exchange of genetic material.

Many earlier microbiologists had presumed that bacteria did indeed reproduce sexually. Van Leeuwenhoek, the very first observer of bacteria as well as many eukaryotic 'animalcules', thought he saw 'Little animals entangled together . . . a-copulating' (1692, in Dobell 1932: 200). But as microscopes improved, such observations were increasingly dismissed as improbable in bacteria. Cohn, the bacterial taxonomist who featured in Chapter 3, thought the testimony of earlier and contemporary microbiologists about sexual phenomena was merely wishful thinking: 'The bacteria have not revealed as yet any true reproduction . . . bacteria only reproduce by vegetative reproduction, not sexually' (Cohn 1875, in Brock 1961: 211). This position had become genetic and evolutionary orthodoxy by the 1940s (e.g., Dobzhanksy 1941): researchers who imagined the possibility of bacterial genetics report being regarded as 'lunatics' (Cavalli-Sforza 1992: 635).

---

[2] I noted in Chapter 3 that the same work refuted (as far as the experimenters were concerned) any role for Lamarckian processes in microorganismal inheritance.

Undaunted and always resolutely pragmatic, Lederberg speculated in the 1940s that establishing microbial genetics was simply a matter of effort: 'How to merge microbiology, genetics and molecular chemistry? ... let's see if [bacteria] have a genetics', is how he retrospectively recalls the situation (Lederberg 1993, in Bivins 2000: 116). While several contemporaneous reviews had collated the lack of success in finding bacterial sex, Lederberg saw these failures merely as encouragement for continuing to search for sexual processes. There is a well-documented story of his extraordinary luck in happening to pick microorganisms that were able to conjugate in an apparently sexual way (Lederberg 1987; Tatum 1959). Lederberg and his collaborator, Edward Tatum (1909–1975), used the K12 strain of *E. coli* (then called *Bacterium coli*[3]) because it would grow on synthetic media, unlike earlier candidates for bacterial proto-genetics. These earlier model prokaryotes included various strains of the respiratory pathogen pneumococcus (now called *Streptococcus pneumoniae*). Although these previously popular organisms did have a process for acquiring genetic material – by transformation, which involves incorporating DNA found outside the cell – working with them was difficult because of their fussy culture requirements.

By using two strains of double mutant *E. coli*, each of which was unable to synthesize certain chemicals, Lederberg and Tatum knew that specific metabolic changes in their organisms could not be due to simple reversion to the non-mutant form. The new characteristics of the organisms on their plates could *only* be interpreted as the result of recombination between strains (Lederberg and Tatum 1946). He and Tatum called this apparently sexual process 'conjugation' although they had no idea of the physiological mechanisms causing it. They thought it might involve total fusion of both the cells engaged in the transfer process. The most immediate conceptual and practical implications of finding recombinational capacities were that bacteria could now be understood as sexual creatures, and crossing experiments could be conducted with them (Hayes 1953). Bacterial genetics was thus officially established. Genetic recombination was soon detected in phages (Delbrück and Bailey 1946; Hershey 1946), and major reviews of the entire field of microbial genetics – including the genetics of established eukaryotic microbes – upheld a continuity of biological process across all life forms (Lederberg 1948).

---

[3] See Mellon (1925) for a claim well before Lederberg that *B. coli* are able to conjugate. It is possible that Mellon's findings influenced Lederberg's choice of model organism, and thus made it less of a serendipitous decision than usually recounted.

Although Lederberg recognized that sexual recombination is only occasional in bacterial populations, this awareness did not affect the big picture he drew of a universal genetic study of life. The main difference Lederberg saw between molecular macroorganismal and microorganismal genetics concerned their scientific targets. He thought that macroorganismal genetics was more likely to focus on mechanisms of hereditary transmission whereas microorganismal genetics was most concerned with mechanisms of genetic variation (Lederberg 1948).

In this early era of molecular genetics, studies of bacterial sexuality proliferated as correspondences between macrobes and microbes were sought and putatively found. Sexual metaphors ran amok as *E. coli* became gendered. Donor cells were conceptualized as male while recipients became passive females; conjugation was referred to as a fertilization process between gametes, and recombined cells called zygotes (Bivins 2000). Microbial geneticist William Hayes (1913–1994), who developed much of the research revealing the mechanisms and consequences of conjugation, almost provocatively deployed the language of animal sexuality while simultaneously revealing the disanalogies between bacterial and eukaryotic sex. These differences included recombination of DNA fragments as opposed to whole genomes, one-way donation rather than mutual exchange, and inefficient mechanisms in regard to the population. Terms Hayes, Lederberg and others managed to use in their publications include ménage à trois, fertility, promiscuity, intimacy, conjugal union, coitus interruptus, penetration and male sex organ (Bivins 2000; Hayes 1966). This use of language shows how a strong commitment to finding the basic, shared properties of life can produce exaggerations of unifying characteristics. These excesses are similar to Ehrenberg's nineteenth-century claims – eventually much criticized – that all the physiological systems possessed by animals were present in infusoria.

Lederberg's discovery with Norton Zinder (1928–2012) of transduction, the process whereby phage and other viruses transfer exogenous DNA into the genome of a cellular host, completed the discoveries of mechanisms of gene transfer (Zinder and Lederberg 1952; Figure 6.1). As the transition to molecular eukaryote biology occurred in the 1950s, some significant differences became apparent between the genomic and genetic organization of prokaryotes and eukaryotes. For example, operons are found in only a few eukaryotes, and only certain kinds of non-protein coding sequence (introns) are found in prokaryotes. Diverse gene regulatory elements were discovered in eukaryotes, and those findings led to the displacement of the operon model of gene regulation that had

Figure 6.1: Major figures in early molecular bacterial genetics: Lederberg, Tatum, Delbrück, Luria, Hayes, Zinder. The Lederberg image is used with permission from the University of Wisconsin (Madison) Archives; Tatum courtesy of www.nobelprize.org; Luria courtesy of profiles.nlm.nih.gov; Hayes used with permission from Australian National University Archives; Zinder used with permission from Cold Spring Harbor Laboratory Archives.

been generalized from *E. coli* (Müller-Hill 1996). Differences between prokaryotes and eukaryotes in numerous other genetic and cellular processes made it appear that there were limitations to the treatment of microbes as model systems. Many molecular biologists as they were now known turned away from viral or unicellular systems to multicellular ones (e.g., Seymour Benzer, Sydney Brenner). In the 1970s and 1980s, microbial population genetics began to make considerable headway as it focused on how clonality and recombination structured microbial populations in ways that were different from macroorganismal popula-tions (Levin 1986). But these developments occurred as population genetics for macroorganisms was transferred to the microbial world, to see if prokaryotic life in particular did have genuine population genetics (Maynard Smith 1995).

Paradoxically, this diminishment of molecular-biological interest in microbial systems as models for macrobial systems was further enabled by the inventions of microbe-derived tools that became central to molecular biology in the 1970s. In Chapter 4, I mentioned Brock's discovery of *T. aquaticus* and how in the 1980s one of its enzymes became central to the in vitro PCR amplification of DNA. Recombinant DNA technology, the major molecular innovation of the 1970s, was built on the basis of microbial capacities to cleave, recombine, clone and express DNA in cells. This set of techniques allowed DNA sequence from any living thing to be genetically engineered. Microbes (or parts of them) thus became technological devices that were useful for how they could be put to work in investigating and manipulating non-microbial biology. In addition, engineered and non-engineered microbes became increasingly important sources of products such as enzymes, antibiotics and bioremediation processes. But very quickly, the new research technologies enabled by these microbial tools brought microorganisms back to being a focus of study in their own right.

A 'new' wave of molecular biology emerged as whole-genome based analyses began to supplement and sometimes substitute for standard genetic methods. A variety of phage genomes was sequenced from 1977 onwards, and whole microbial genomes – bacterial, yeast, archaeal – capitulated to sequencing efforts in the mid-1990s. The previous chapter on microbial ecology showed how genomics eventually encompassed 'wild' as well as domesticated laboratory microbes. Prokaryotes now became the focus of multiple lines of genome-based inquiry, due once again to their tractability (that is, being easier to sequence, because of having smaller and tidier genomes than eukaryotes) as well as their putative capacity to represent all life at the molecular level. Yeast, as a eukaryotic microbe in the group (fungi) most closely related to animals, also consolidated its hold on being a eukaryotic model system for genetic and now genomic analyses. Yeast researchers argued that combining genetic and genomic approaches with ecological and evolutionary research on yeast would 'create the first "general" model organism in biology: a complete, defined system, suitable for study from any perspective' (Replansky et al. 2008: 500).

Genomics, while it might have been considered a distinct methodology or even a field in the 1990s and early 2000s, quickly became perceived as simply a technology for generating data. Making better scientific sense of that data became the mission of systems biology, which is the methodological solution to the problem of integrating large-scale molecular data, experimental hypothesis testing and mathematical modelling of molecular dynamics (O'Malley and Dupré 2005). But just as was the case for earlier molecular biology and genomics, the material basis of systems biology is found in microorganisms. Most systems biology is network-based and focuses on explaining and predicting the dynamic processes of gene-regulatory, signalling and metabolic networks. Achievements so far have been to construct mechanistic models of individual genome-scale networks, which link genotype-phenotype maps of microbes, particularly prokaryotes. Collectively, these genome-scale networks can be pieced together to create comprehensive virtual cell models – in the most complete case so far, of a very small bacterium, *Mycoplasma genitalium* (Karr et al. 2012).[4] One reason for this microbial focus is that networks in eukaryotes are too extensive and complicated, and the genomic bases of these networks are far less well known. Modelling eukaryotic networks thus becomes unfeasible, for the time being at least.

Synthetic biology, which is a developing set of methods to make material models of biological processes with the aid of mathematical 'design' models, is also primarily carried out in model microbial systems (O'Malley et al. 2008). Just one example, with important evolutionary implications, is a synthetic construct that shows how it is possible to engineer altruistic 'death circuits' into *E. coli* in order to find out the conditions in which programmed cell death is advantageous to a population (Tanouchi et al. 2012). Considerable efforts are being made to transfer microbe-based models and constructions to multicellular eukaryotes (or even unicellular eukaryotes), but progress is slower in these systems (Purnick and Weiss 2010). However, some headway is being made on engineering interactions between different microorganisms, with the aim of using them as model systems for the study of intercellular interactions in general (e.g., Wintermute and Silver 2010). Some of these microbe–microbe interactions are analogous to cell-to-cell communication processes in multicellular organisms, and may permit studies of evolutionary transitions in which communicative capacities are thought to have played important roles. All these molecular approaches – genomics, systems biology, synthetic biology, and molecular biology in general – presume that general features of molecular structure and function, reproduction, variation, heritability and fitness obtain in all forms of life. These approaches also assume that the general properties of living things can be most effectively studied in

---

[4] *M. genitalium* as a model organism for systems biology has proved to be something of a problem. Although it has a very small genome, which is handy for sequencing and genome minimalization, this bacterium is not very experimentally tractable.

microorganisms. Building on similar assumptions has been a major aspect of the justification of experimental approaches to evolution and ecology.

Experimental evolution, ecology and developmental biology in microbial model systems

Evolutionary studies have a long history of referring to microbes to make basic claims. An early use of microbes as the objects of evolutionary analysis occurred as a generation of scientists attempted to find experimental evidence for and against Darwinian evolution. Darwin himself had argued that microbes fell within the remit of his evolutionary theory (1861). He made at least three major evolutionary claims about microbes (O'Malley 2009). The first was that all organisms, no matter how different they seem to be from animals and plants, undergo natural selection. His second point was that microbes demonstrate by their very morphology that evolution cannot be conceived as a progressive process that moves from simple to complex. His third argument was that microbes and their adaptive capacities are very important biological phenomena to examine if the history of evolution on Earth is to be elucidated. Although little recognition has been accorded to Darwin's discussions of microbial evolution (O'Malley 2009), it would have been very peculiar if he had not thought that microbes fell within the scope of evolution by natural selection. Some of his contemporaries went further, however, in trying to demonstrate or disprove these claims.

James Samuelson (1829–1918) was a natural historian, microscopist and popular science writer who wrote broadly about biological matters. He carried out experiments in the 1860s to demonstrate how a single monadic form of life could develop into all the varieties of microscopic life, from simple to increasingly complex. This developmental theory, he argued to the Royal Society and Darwin, did away with any need for the 'metaphysics' of natural selection (Samuelson 1865). Simultaneously these experiments were deemed to have refuted spontaneous generation, which was a major worry for Darwin's contemporaries. On the supporting side for Darwin's theory was some basic experimentation on the evolvability and adaptability of bacteria. William Dallinger (1839–1909), a cleric and microscopist, carried out seven years of continuous culture experiments on three types of 'monad' in increasingly warm environments. His results could be interpreted as adaptive evolution in response to rising temperatures. This research, he believed, would 'palpably demonstrate [Darwin's] great doctrine' (Dallinger 1878: unpaginated). Darwin thought these experiments were intriguing, but because the theoretical battle was being fought over organisms closer to humans, he never drew on Dallinger's results.[5]

Although the middle decades of the twentieth century saw many bacteriologists testing evolutionary hypotheses on microorganisms (including viruses),[6] the name many people associate now with 'experimental evolution' is that of Richard Lenski. His research from 1988 until now on tens of thousands of generations of *E. coli* has been the source of many profound insights into evolutionary process and theory. Very seldom is it qualified as providing 'merely prokaryotic insight'. It has generated better understanding of evolvability, hypermutation, the interplay of neutral and adaptive evolutionary processes, convergence and kin selection (Elena and Lenski 2003; Buckling et al. 2009). Populations of *E. coli* can be frozen and brought back to life; new environments and adaptive possibilities can be forced on samples; predators and sexual proclivities can be introduced (Lenski et al. 1991). Entirely novel capacities have evolved in populations with particular genetic backgrounds. The most dramatic adaptation observed so far is the capability to consume citrate aerobically, which *E. coli* are normally unable to do in oxic conditions (Blount et al. 2008). Although species definitions are problematic for prokaryotes, as Chapter 3 discussed, this newly generated capacity is interpreted as part of a speciation event. Overall, Lenski's long-running experiment shows great potential for bridging the gap between microevolution and macroevolution (Lenski and Travisano 1994), and has inspired and contributed to a huge range of additional experimental and other work.

One of the main attractions of experimental evolution is that rather than relying on the comparative method and historical narratives, it can test causal hypotheses about evolutionary scenarios much more directly. These scenarios include models of evolutionary transitions (Celiker and Gore 2013). One evolutionary transition that is of enormous biological and philosophical interest is the transition to multicellularity. It is misleading to speak of it as a singular event, however, because multicellularity has evolved independently many times, with new findings of multicellular forms being made every year (e.g., Brown et al. 2012; Grosberg and Strathmann 2007). The common factor, however, and one that goes without saying, is that all these transitions began with a unicellular organism that evolved some sort of multicellular

---

[5] In addition, says evolutionary biologist Graham Bell (2010), prevailing sentiments at the time that evolution should happen very gradually, and not in the lifetime of one observer, meant that Dallinger's work was seen as misguided.

[6] See Levin (1981) for a synopsis of population genetics in *E. coli* from the 1950s to the 1980s. Some of this research is referred to in the discussion of chemostat research below.

state and not necessarily an obligate one. Chapter 4 mentioned some forms of multicellularity, but did not discuss just how many have been thrown up by evolution. They include colonial, filamentous, aggregative and 'complex' forms (Knoll 2011; Brown et al. 2012), none of which need have any immediate fitness advantages (Niklas and Newman 2013). It is apparent, however, that there is something about being a eukaryote that makes multicellularity a viable and even a likely option (O'Malley et al. 2013).[7] With this in mind, it seems reasonable to think that experimentally, multicellularity could evolve in a population of unicellular eukaryotes – given time and appropriate conditions.

It is this very reasoning that inspired a recent attempt by evolutionary biologists Michael Travisano, William Ratcliff and colleagues to evolve multicellularity in vitro (Ratcliff et al. 2012). Using yeast, which has an ability to clump together and settle in a liquid medium, the experimenters selected 'settlers' and got rid of the yeast that settled less. Multicellular 'snowflake' phenotypes of yeast 'evolved' in just a week, in the form of adherent cells (not just aggregations) that were the results of generations of mother and daughter cells remaining attached to one another after fission. To reproduce, one cell would break off from the multicellular organism and start its own multicellular development, thus going through a unicellular bottleneck. The multicellular forms undergo developmental processes that have many similarities to metazoan multicellular processes (Ratcliff et al. 2012; see below for further discussion of development). Naturally, there are limitations to what this study shows about any transition from unicellularity to multicellularity. Selection for settling, as the authors note, 'was chosen not because it is widespread in nature, but rather because it is an experimentally tractable method to select for larger size' (Ratcliff et al. 2012: 1595). In addition, fungi, whether unicellular or multicellular, usually are not regarded as the prototypic eukaryote despite their phylogenetic closeness to metazoans. Although numerous fungi are multicellular, there is also evidence that some lineages have reverted to unicellular states from multicellular ones – not something that occurs in obligately multicellular animals (Parfrey and Lahr 2013). However, the success of yeast as a model eukaryote for genetics and genomics – and in other experiments inducing multicellularity (e.g., Koschwanez et al. 2011) – probably overrides phylogenetic considerations such as these. As philosophers Arnon Levy and

---

[7] And multicellularity, as noted already (see Chapters 4 and 5), can be very effectively studied in prokaryotes. The highly social bacterium, *Myxococcus*, is the exemplar organism for such studies (Velicer and Vos 2009).

Adrian Currie (2014) suggest, this specific experiment with yeast involves modelling in its strongest sense, when the organism is used as a surrogate system for a general process, rather than to make empirical extrapolations to related organisms.

This surrogacy role, in which microorganisms model general processes, comes out very clearly in ecological research done with microbial model systems. Ecological experimentation on microbes has had a major impact on general ecological and evolutionary theory (Jessup et al. 2004).[8] This tradition probably is most famously exemplified in Georgy Gause's (1910–1986) experiments with microbial eukaryotes – *Paramecium* and yeast in particular. His experimental and mathematical modelling of competitive dynamics between microbial species led to the 'competitive exclusion principle', which dictates that new niches must be constructed if different taxa are to share the same resources (Gause 1932; 1934).[9] Some exemplary ecological experiments have subsequently shown how this principle can not be straightforwardly true, and that it requires additional explanation to understand the experimental equilibrium that can be achieved between two species on a single resource (e.g., Levin 1972).

The invention of the chemostat in 1950 was a boon to laboratory-based ecological experimentation on microbes.[10] A chemostat is a device in which a culture medium is steadily added to a contained culturing environment, usually with one nutrient in the medium provided at a lower rate than would be required by a maximally dense population of microorganisms. Old cells and spent media are removed at the same rate so that volume is kept constant (see Figure 6.2). The culture is continuous and can be described formally by kinetic growth equations (Dykhuizen 1990; Ferenci 2008). Chemostat experiments have contributed enormously to ecological and evolutionary understandings of microbial populations. Even short-term chemostat experiments have evolutionary consequences, often with genetic explanations (Dykhuizen and Hartl 1983). Ecological chemostat experiments have revised earlier findings such as those that denied the stable coexistence of predator-prey relationships in homogeneous environments (Chao et al. 1977). Numerous other studies have showed that instead of competitive

---

[8] See Jessup et al. (2005) for many examples of important twentieth century experimental microbial ecology.

[9] See Hardin (1960) for the numerous historical and philosophical complexities of competitive exclusion – a label he bestowed on the process.

[10] The chemostat was in fact invented twice in 1950, by biophysicists Aaron Novick and Leo Szilard, and by geneticist Jacques Monod. Monod's term for the device – 'bactogène' – did not catch on, but his equations for the growth of populations within it did.

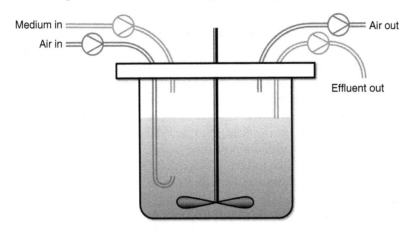

Figure 6.2: Schematic view of a chemostat. For more detailed descriptions and depictions of the machinery involved, see the figures in Dykhuizen (1993).

exclusion being the outcome, distinct genetic variants (polymorphisms) can arise within populations and stably co-exist (e.g., Rosenzweig et al. 1994). Even in the very constant environments provided by chemostats, bacteria construct new niches based on mutations, thus enabling populations to divide into interacting strains and flout the competitive exclusion principle (Ferenci 2008).

I mentioned marine food webs in Chapter 5, and how a better understanding of microbial interactions had overturned old ideas of a simple chain of consumption from small to large organisms. More recent investigations of ecological complexity in microbial systems extend these findings. For example, it might be expected that an artificial ecosystem with three microbial taxa (an alga, an *E. coli* and a ciliate protist) is a simple one, with obvious roles in the food web for each organism of (respectively) primary producer, detritus consumer and predatory consumer. However, in one very illuminating experiment, the ecosystem rapidly subdivided into different niches in which the organisms generated diverse and even unusual capabilities, or lost them (Hekstra and Leibler 2012). Relationships between the organisms varied greatly in all the niches, and massively increased the expected biological capacity of these fairly artificial ecosystems. The population dynamics of the organisms in these experiments can be modelled statistically and the average dynamics generate basic quantitative principles (Hekstra and Leibler 2012). Artificial ecosystem experiments based on microbes are likewise highly revealing of basic but very general processes, such as multilevel selection (Swenson et al. 2000; see Chapter 4).

Despite the generation of rich insights, there are many caveats associated with chemostat use that tie into questions about model systems in general. As Jannasch noted, despite its experimental usefulness, the chemostat 'represents an utterly unnatural situation. It is neither meant to reproduce nor [is] capable of reproducing a natural habitat. Its sole purpose is to make physiology, biochemistry, and genetic responses of whole populations amenable to arithmetic analysis' (Jannasch 1974: 717). And even with such 'unnaturalness' taken into account, there are additional confounding factors that spoil the very simplification of the system: 'Despite the experimental constancy it is impossible to keep biological variation out of consideration when dealing with chemostat cultures as with any other living system' (Ferenci 2008: 215).

Another major caveat to take into account, in light of the previous chapter on microbial ecology, is that microbial experiments in the laboratory are necessarily done on cultured organisms and whatever the extent of culturability – it doesn't matter if it is less than one per cent or 50 per cent – many microorganisms will not be represented in these experimental studies. Due to culturing constraints, laboratory-based studies of model systems, such as the *Rhizobium*-legume mutualism discussed in Chapter 4, have produced excessively simplified understandings of the mutualism's dynamics (Dini-Andreote and van Elas 2013). Although field experiments do exist, in which natural microbial communities are manipulated in various ways, experimentation in 'wild' systems is much less common than in laboratory-based cultures. This is at least partly because of the difficulties in setting boundaries for naturally occurring microbial communities.

Tests of ecological and evolutionary assumptions have sometimes been carried out with specific claims about how microbial systems function as experimental models. A philosophically intriguing one, from a study exploring whether generalists are less efficient consumers than specialists, is that model system experiments serve similar epistemic purposes to those of mathematical models: they are both 'heuristics', 'to show what is possible' (Dykhuizen and Davies 1980: 1213). However, in the case of the experimental microbial system, there is a major advantage over the mathematical model. Because the organisms produce 'unanticipated phenomena' in short time-spans, these experimental systems can function as 'an intermediate between mathematical theory and field observations' (Dykhuizen and Davies 1980: 1213; see also Jessup et al. 2005: 283–284; Gause 1934: 10–11). Mostly, however, microbial model systems are chosen in ecology and evolution for conceptually unadorned practical reasons. Although long-term experimental evolution and ecology can be done with animals and plants too – and even digital 'life forms' – there are overwhelming advantages of convenience and realisticness on the side of microbial populations and communities (Kawecki et al. 2012). Experimenters can control many aspects of the model microorganisms and their environments, and replication can be done repeatedly (Jessup et al. 2005). In addition, researchers can revisit previous experiments in frozen cultures.

Despite the many differences between microorganisms and macroorganisms, experimenters usually argue that ongoing experimentation will help sort out 'general principles' of life from 'idiosyncratic features' of specific lineages (e.g., Kawecki et al. 2012: 552; Andrews 1991). Not everyone is convinced. Generalizing from microbial systems to other ecological and evolutionary systems is frequently argued against when general principles are sought. These reasons are similar to those discussed in the previous three chapters for making a major division between unicellular and multicellular life (or prokaryotes and eukaryotes): the scale, organizational structure, population sizes and reproductive mechanisms are just too different in microorganisms. Some of the hesitation in using microbial model systems, however, has more to do with the nature of experimental systems in general rather than microbial systems in particular. Evolutionary ecologist (previously physiologist, lately anthropologist) Jared Diamond, for example, thinks 'bottle' experiments such as Gause's are 'uselessly unrealistic': 'It is not even possible to know what variables are worth putting into one's bottles until their importance is established by other means' (Diamond 1983: 586). This claim seems rather extreme, however, because not all important factors are generally known in advance of inquiry, and every form of scientific investigation requires simplification at various points in the research process. Oversimplifications are not the problem exclusively of experimental ecology, microbial or otherwise, but can be found in observational studies, mathematical modelling and conceptual theory.

Many experimental evolutionary ecologists would argue against criticisms of experimental microbial systems along these lines:

> Experimental evolution does not seek to mimic specific systems in all their complexity. Instead, the microcosms are biological models in which researchers attempt to capture the essence of evolving systems in general [to] shed light on general processes that are expected to occur in all life on Earth (Buckling et al. 2009: 827).

Moreover, as Diamond admits later in his opinion piece above (1983), robust results are usually found by using a variety of methods. Investigating general ecological or evolutionary claims in microbial populations and communities is unlikely to be the sole basis of claims made about life in general. The question this chapter is asking is not primarily one of whether these experiments are simplifications (of course they are) but about whether microorganismal systems are reliably representative of living systems in general. One way in which to think more about this question is by looking at microbes as model systems for development.

The very idea of microbes as models for ontogeny could be quite a surprise for developmental biologists, who usually think that development is exclusively a process undergone by multicellular organisms, most distinctly in metazoans (see Chapter 4 for a discussion of microbial development). These tensions are very helpfully addressed in a discussion proposing the use of microbes as model systems for development. Travisano and philosopher Alan Love show how microbes can be understood as models for specific manipulative and representational purposes in a multilevel research strategy (Love and Travisano 2013). By multilevel, they mean the way in which modelling moves from a microbial model to a metazoan one. Microbial models of development have their limitations, but so do all model systems – hence the need for integrative biological practice to find and fill the gaps.

To make this argument, Love and Travisano distinguish between *developmental phenomena* and *developmental mechanisms*. Many microbes, eukaryotic and prokaryotic, are very useful model systems for developmental phenomena such as aggregation, adhesion, cellular division of labour, and intercellular communication and coordination. Some of the mechanisms underlying those developmental phenomena are also very similar to those of metazoans (for example, signalling networks). Representation at the level of phenomena, however, is not necessarily accompanied by representation at the mechanism level (and no doubt vice-versa). But the other criterion of model system choice is manipulation, which is at least as often the dominant reason for modelling developmental processes on the basis of microorganismal experimental systems. However, Love and Travisano emphasize that even the well-justified use of microbes as model systems for development and other multicellular processes does not mean the replacement of animal and other models. Microbial models are simply 'a strategic part of a more comprehensive integrative strategy' (Love and Travisano 2013: 165). A research programme with the goal of being more comprehensive is also likely to include studies that examine more deeply how microbes are implicated in animal and other large-organism development (see Chapter 4). Understanding how these integrative strategies work is an important part of understanding the role microorganisms play in producing general understandings of life.

Representation and tractability in microbial model systems

So far, the discussion has pointed to the main reasons that are given for using microbial systems to examine and model phenomena that occur in large organisms. One is that they can adequately represent basic biological features, and sometimes even more complex and specific ones. The other is that they are highly tractable experimental systems that can be manipulated easily, and the findings used to fulfil representational purposes. Combined, these reasons contribute to what might be understood as a general 'unity of life' thesis that makes claims about biological entities as well as scientific practice. How this thesis works in light of microbial knowledge has important implications for philosophical studies of the life sciences.

Microbial systems, as Love and Travisano (2013) emphasize, are not meant to be full and perfect representations of multicellular systems. Model systems are chosen to represent particular phenomena and those choices are underpinned by representational ideals, involving the inclusion of the most crucial causal factors but not the peripheral details (see Weisberg 2007). The representation achieved by model systems always, therefore, involves simplification and incompleteness – models are never facsimiles of the phenomena to be modelled, whether microbial or macrobial. As microbial ecologists themselves argue,

> Laboratory model systems [of microbes] are not intended to be miniature versions of field systems, and laboratory ecologists do not intend to reproduce nature in a laboratory model system. Rather, the purpose of laboratory model systems is to simplify nature so that it can be more easily understood (Jessup et al. 2004: 189–191).

There are different rationales underpinning this simplification process, however. Developmental biologist Jessica Bolker (2009) distinguishes between exemplary representation and surrogate representation, with the former being models specifically targeting a much broader phenomenon, and the latter more of a proxy – an indirect means for getting at features that are otherwise difficult to access. It could be argued that microbes are 'merely' surrogate models that are not intended to be very generalizable, but the very justification of exemplary model organisms is that they have 'provided the foundations of most of our understanding of biology, and allowed us to grasp and articulate its broadest principles' (Bolker 2009: 489). This seems to be the case for many of the uses of microbial systems in molecular biology, evolution and ecology. They are used to represent phenomena other than themselves; they are generalizable abstractions of specific aspects of biological complexity; they are tractable because of an established history of use and scientific familiarity. Model systems thus play broader roles than model organisms in many cases (Ankeny and Leonelli 2011), and can also incorporate the more specific function of 'tools' that microorganisms perform (see above).

There are other less straightforward ways in which microbial systems have been used to represent macrobial phenomena. These occur when the use of microbes as model systems for certain metazoan phenomena and mechanisms might seem incongruous, because microbes are thought *not* to have these traits. For example, intercellular communication and cooperation have been suggested as inappropriate phenomena to be sought in microbial model systems (e.g., Bolker 2009: 493). However, it is clear that these phenomena *do* exist in microorganisms (as the previous chapters have pointed out and I point out again, below). This sort of concern is probably grounded in a more basic claim that microbes are so extremely simple that they couldn't possibly represent the complex features of plants or animals (e.g., Ankeny and Leonelli 2011: 320). Sometimes, however, the very use of microbes as models of general life processes reveals unexpected similarities between microbial and macrobial systems. Making surprising analogies, sometimes without an initially solid substantive basis, is a prominent discovery mechanism in numerous fields of microbiological investigation.

For instance, bacterial chemotaxis, which is cell movement enabled by responses to environmental chemicals (see Glossary), began its molecular career rather ambitiously as the study of 'sensory behaviour' or, even more radically, as a 'molecular approach to neuroscience' (Hazelbauer 2012: 295). Early molecular researchers presumed there were analogues of recognition, response and memory at play in microorganisms – although often these were more metaphorical claims than strict hypotheses (Berg 1975). As suggested by Julius Adler, an early and important molecular chemotaxis researcher and another graduate of van Niel's general microbiology course,

> Modern studies of biology have revealed a universality among living things. For example, all organisms have much in common when it comes to their metabolism and genetics. Is it not possible that all organisms also share common mechanisms for responding to stimuli by movement? ... from this point of view one may hope that a knowledge of the mechanisms of

motility and chemotaxis in bacteria might contribute to our understanding of neurobiology and psychology (Adler 1966: 715).[11]

The field of chemotaxis studies is now huge and multidisciplinary, and one of the reasons it has grown so much is because it does indeed work as a 'model behavioural system', from which it is 'possible to gain insights into the human brain' (Koshland 1980). This is achieved by emphasizing 'those features of the bacterial system that are relevant to all species and provide windows into the interpretation of more complex phenomena' (Koshland 1980: vii-viii).

Social evolution is another good example. Originally studied only in animals because of a perceived need for sophisticated kin recognition systems, social evolution research is now flourishing on the basis of microbial model systems (Foster et al. 2006; Velicer 2003; see Chapter 4). Not only have microorganisms, including prokaryotes, been found to manifest complex social behaviours that have evolutionary ramifications, but also their genetic bases can be studied much more effectively in such systems. Genetic mechanisms underpinning behaviours associated with kin recognition, cheater suppression and mutualism have been a gold mine of knowledge for phylogenetically broad understandings of social evolution (Foster et al. 2006; Velicer and Vos 2009; Celiker and Gore 2013).

Research that is perhaps even more surprising occurs when microbes are used as model systems for cognition and communication. Numerous microorganisms, especially protists, exhibit behaviours that imply memory and a capacity for spatial mapping (e.g., Reid et al. 2012). Prokaryotes and other microbes exhibit a range of what are often called 'cellular decision making' processes (Balázsi et al. 2011: 910), which depend on a variety of sophisticated mechanisms. They may also be 'irrational' in these choices, if absolute preferential ranking is assumed. If, however, choices are comparative – dependent on what is available – these decisions by micro- and other organisms can be understood as rational. There is some evidence that the slime mould *Physarum polycephalum* uses 'the same comparative decision-making processes as do neurologically sophisticated organisms' (Latty and Beekman 2010: 312). Both individually and as groups, prokaryotes can sense their environments and respond with different behaviours. These behaviours are increasingly understood as non-deterministic, and

---

[11] Considerably later (in the 1990s), Adler (2011) admitted that while the phenomena of behaviour are similar in bacteria and large organisms, the mechanisms are very different (see Love and Travisano 2013 and above for the distinction). Nevertheless these differences were found by assuming similarity.

even argued to be manifestations of a 'fundamental biological property' that is very effectively studied in microbes but can be generalized to many organismal behaviour and decision-making strategies (Balázsi et al. 2011: 922).

Animal communication is often analogized to quorum sensing in microorganisms. Quorum sensing is a population-level mechanism that works by prokaryotes responding to the density of other prokaryotes via small signalling molecules. These molecules regulate the gene expression of various capacities, including movement, changes in cell state, biofilm development, DNA transfer and acquisition, and symbiotic interactions that involve the sharing of public goods (Darch et al. 2012). In virtue of how it works in populations, quorum sensing is often deemed to enable prokaryotes to 'act as multicellular organisms' (Waters and Bassler 2005: 319). This capacity makes quorum-sensing groups excellent model systems for understanding cheaters and social conflict (Diggle et al. 2007). The cost of signalling means for some social evolutionists that inter-species quorum sensing should be understood as coercion or manipulation instead, since it cannot be explained by kin selection (Keller and Surette 2006). But the very fact that at least some of these 'communicative' interactions appear to produce mutual benefits (Atkinson and Williams 2009) makes quorum-sensing a highly effective experimental system for the evolution of communication. Quorum sensing may even illuminate the apparently complex phenomenon of anticipation. This is imputed to situations where bacteria appear to make decisions, such as whether to sporulate or enter stationary phase, based on predictions about future environments from current environmental cues (Goo et al. 2012).

One reason why microbes are such good model systems is that they are living entities that always do more than the designated modelling task. They do not have arms, legs, livers or stamens, but they 'say' more about biology as research is done with them – even if they do not say exactly what is required to allow multicellular organisms to be paradigm organisms. But for most research, representation is trumped by tractability when it comes to choices about model systems, microbial or otherwise (Love and Travisano 2013). Tractability, which includes manipulability, is one of the main justifications given for turning to microbes, because not only can these organisms be 'controlled' – in the pure culture sense of knowing exactly what is there, assuming clonality and no recombinational differences in a population – but the environments can be, too. Complex nutritional requirements can be done away with for laboratory microbes, and a simplified system created and studied very effectively (Rainey et al. 2000).

Some of these decomplexifications are very artificial – 'simple' communities, for example, with only two or three interacting taxa – but others are more realistic, such as particular predator-prey relationships.

Manipulation is a crucial factor in the choice of model systems. Because numerous experimental interventions at different biological scales can be conducted in microbial systems, causal inferences can be pushed further across diverse systems of interaction and phylogenies (Rainey et al. 2000). Moreover, conceptual manipulation is a capacity of some importance in these systems as well. I mean this in two senses: one in the expected sense of such systems providing empirical sounding boards for conceptual models; the other in the sense of enabling philosophical theory building. In regard to the first, there are numerous examples above of how microbial model systems have been immensely useful for conceptual interventions when, for example, questions are asked about how competitive exclusion might work and when it won't, and what introducing niche construction will do to a broad ecological theory such as competitive exclusion. In reliance on broader knowledge about microbial capacities for phenotypic plasticity, mutation and niche reconstruction, conceptual claims about evolutionary relationships can be re-examined and made more fine-grained. This can of course occur with many other organismal systems, but more detailed, lengthy and comparative studies have been done with microbes and they thus have more conceptual force.

Where conceptual manipulability really pays off for philosophers, however, is in the second sense. Gaining in-depth understanding of general biological phenomena – such as reproduction, speciation, individuality, sociality, cooperation, biodiversity and so on – of course can be done on the model system basis of zebras or honeybees (and often has been in philosophy of biology). However, doing so will produce a very narrow conceptual world in which to carry out philosophical work. By using microbes as model systems, philosophers can manipulate concepts in much more interesting, broad and demanding ways, in order to see the consequences of particular interpretations or modelling strategies, and to make the best philosophical and scientific sense of knowledge-making frameworks. After having reflected on microbes, philosophers can default to their favourite mammal again, but these large-organism applications will then be at least adequately contextualized (that is, in relation to microbes).

The teleosemantic programme of philosophical research discussed in Chapter 1 illustrates how a microorganismal model system has been used to develop a philosophical theory. Magnetotactic bacteria, even if they are not doing what Millikan, Dretske and others thought they were, nevertheless allow the very notion of representation to be thoroughly examined, compared in a variety of instances, and a general concept to be clarified by distinguishing between bacterial and human representational activities. This is in fact what Millikan did when she specified what magnetotactic bacteria could *not* do representationally vis-à-vis human capacities for mental content and intentionality (1989; Chapter 1). Other projects in the philosophy of cognition and elsewhere have similarly benefited from using microorganismal capacities as basic comparators for philosophical analyses of human abilities. By beginning with microorganisms responding to environments and processing information, for example, philosophers can develop a gradient of evolved capacities that culminate in the human ones (e.g., Godfrey-Smith 2002; Skyrms 2010). However, there are limits to how far such 'culmination' approaches can be taken, as Chapter 1 discussed and will be elaborated further below.

A general message to be drawn from how microbial model systems are used in biology is simply that for many biological research purposes, there are *not* two fundamentally different types of organism, with prokaryotes set apart from eukaryotes (or microbes from macrobes). For most microbial ecologists and evolutionists, the same major biological imperatives face microbes as much as macrobes: survival, maturation, coping with changing environments and interacting in fitness-enhancing ways with other organisms (Andrews 1991; Jessup et al. 2005). Likewise, genetics and molecular biology, from their earliest instantiations to the ones we know today, have a 'unity of life' thesis at their core. Even though microbes were still conceived of as anomalous forms of life for much of the first half of the twentieth century – largely because of not having detectable compartments, such as nuclei, within their cells – their outlier ontological status obviously did not prevent them from becoming the epistemic standard bearers of the great golden era of molecular biology. Viruses, bacteria and eukaryotic microbes were conceived as paradigmatic life forms that could then justifiably be used as reproducible experimental systems and laboratory instruments.

In the case of molecular microbial genetics, this strategy paid off handsomely, despite cautionary reflections such as those of the ciliate geneticist, Tracy Sonneborn (1905–1981):

> In the past, the main findings of genetics have proved to be universal in applicability; this underlies the confidence of many contemporary biologists in the applicability to man [i.e., humans] of the main new findings about microbes. Whether this confidence is justified remains to be seen (Sonneborn 1965: 237).

Despite early suspicion of microorganismal similarities, molecular microbial genetics was built rather incautiously on two assumed dimensions of unity of life. The first refers to living things being unified at a molecular level (and possibly other levels); the other to the associated assumption that the same molecular approaches and generalizations can be applied to any sort of organism, regardless of size or organization. The unified understanding of inheritance, variation and reproduction produced in the 1940s and 1950s of course should have had implications for a unified understanding of evolution, but as we have seen in Chapter 4, the modern synthesis did not encompass microbes in any equal or conceptually important way. This disconnection – which is especially curious given that genetics captures the variation and heritability considered essential material for selection – means that there were perceived limits to how far any assumption about unity of life could be taken.

One question to ask about this situation, which is also a question for how philosophers use microbial model systems, is whether unity-of-life research strategies are as straightforward as thinking of microbes as the basic no-frills system, with the sophisticated biology building on and elaborating such systems. Delbrück, for example, after moving beyond his early ideas about unity being derived from physics, initially thought that microorganisms could be simple models for more complex life. He found this not to be the case:

> The organism [a molecular biologist] is working with is not a particular expression of an ideal organism, but one thread in the infinite web of all living forms, all interrelated and all interdependent ... The student of evolution appreciates this game of segregation and recombination as a clever trick for trying out heredity in new combinations; it is not the basic thing but a refinement and elaboration of the *really* marvellous accomplishment: ordinary, uni-parental reproduction ... The complexities of sexual reproduction and of recombination are not eliminated by going to this seemingly elementary level [of viruses and microorganisms] (Delbrück 2000 [1949]: 9, 13–14).

Delbrück's exclamations about reproduction are of relevance for the discussion in Chapters 4 and 5 about whether multilineal collectives are individuals in any sense. Although they are often failed or made marginal because of not reproducing like animals, Delbrück's observations can be added to the earlier argument against seeing animals as the paradigmatic reproducers. It is hard to imagine molecular biology having got very far if its practitioners had not agreed with Delbrück about the value of understanding microbes as model reproductive systems.

Another interesting aspect of Delbrück's view of microorganisms as models for the life sciences was his observation that the 'unity of interconnectedness' was not the basis on which the traditional unity of science was deemed to rest. The older understanding of unity was commonly conceived as dependent on universal laws, which did not work for molecular biology, he thought.

> Every biological phenomenon is essentially an historical one, one unique situation in the infinite total complex of life. Such a situation from the outset diminishes the hope of understanding any one living thing by itself and the hope of discovering universal laws, the pride and ambition of physicists (Delbrück 2000: 10).

Although today's molecular biology might also not be overly optimistic about laws, researchers still hope to find regularities that have broad purchase in the biological world and can be used to explain and predict many different biological events. I will discuss how we might understand these 'principles' in the concluding chapter, but here I want to think about unity as something achieved on the basis of and in relation to microbes. As many molecular, evolutionary and ecological biologists have discovered, it is increasingly valuable to generalize from microbes to other life forms, and the more it is done, the more 'connected' the biology becomes. This is not due to seeing molecules as the elementary particles of biology, but to the view that microorganisms in all their complexities and organizational plasticity are basic to an understanding of all life.

One unanticipated outcome of the older molecular unity programme, achieved on the back of microbiological research, was in effect the germ of a new philosophy of science (for the biologists if not the philosophers of the time). Far from allowing practitioners to think of physics as the most fundamental science, the unifying capacities of molecular biology suggested the conclusion that understanding life – even at its most microscopic and supposedly least complex level – was not going to be achieved solely on the basis of physical and chemical reductions. Although theoretical reductionism of the old sort has been left behind by philosophy of science, and never greatly perturbed the molecular life sciences anyway, I want to suggest a different sort of shift is still needed in philosophy of science and philosophy of biology in relation to microbiology.

Big-picture philosophy of biology

In many of the previous chapters, the overall message could have been interpreted as 'microbes are equal to macrobes'. But as the book draws to a close, there is a less cautious moral to extract from the microbial story:

microbes are superior to macrobes, for philosophical purposes at least. This proposal may raise concerns. As evolutionary geneticist Bruce Levin remarked, 'I could suggest that what is true for selection in bacterial populations is also true for selection in populations of eukaryotes … But such suggestions would be presumptuous' (Levin 1988: 12).[12]

Although it may indeed be presumptuous to suggest that starting with microorganisms is best, and, from the other direction, that there are major limits to starting with multicellular eukaryotes, these suggestions should not appear to be too unreasonable by now, given everything that has been discussed already in the earlier chapters. To base general claims on organisms that make up many more forms of life over many more millions of years than their later and rarer descendants seems a perfectly sensible strategy.

To make this point more strongly, let me recap the messages from the preceding chapters. I argued in Chapter 2 that microbial classification is the basis for the highest-level classificatory divisions in biology. These are the domains, and they are threefold: Archaea, Bacteria and Eukarya. Humans, other animals, plants and fungi – all the organisms standardly conceived as multicellular – fit into the third group (which may be just a group within Archaea anyway). Whether using phylogenies, cell biology, molecular analyses or a combination of these approaches, even eukaryotes in a broad classificatory sense are microbial, with large organisms being simply special cases of the smaller microbial ones.

Microbial evolution is, without question, concerned with most of the evolution on Earth, and it covers the earliest life forms from around 3.5 billion years ago to current life forms. All existing evolutionary mechanisms evolved in microbes, and most major evolutionary transitions were microbial – especially if transition schemes are not too focused on the wonders of human or other metazoan evolution (Chapter 1). The more recently evolved life forms of standardly conceived multicellular organisms not only emerged from microbes and inherited their evolved capacities, but also evolved in the way they did because of microbial features or interventions. Evolution in these general senses is 'microbial', and even organisms such as ourselves are evolving microbially. We are a twig that has branched within the Opistokonta (see Chapter 2: Figure 2.5), a thoroughly microbial group, and we are coevolving with our symbiont microbiota, as the discussions in the microbial ecology chapter showed.

---

[12] This is the same Levin whose co-authored paper, 'Bacteria are different' (Levin and Bergstrom 2000) was mentioned at the start of Chapter 4.

Microbial ecology itself focuses on the most widespread diversity of life, which is microbial, and the biogeochemical functions that microbes effect to make other forms of life in any ecological context possible. There is no ecosystem that is not at its basis microbial, and no ecological interaction that does not involve microbes. Furthermore, by changing the biogeochemistry of the planet, microbes made possible the existence of larger life forms, such as ourselves, and microbes continue to sustain and regulate those conditions (for example, the oxygenation of the Earth's atmosphere and ocean). All the molecular, biogeochemical and other ecological research I drew on in previous chapters showed how microbially dependent we are, and how understanding any living system requires understanding microbial systems. Microbial consortia, whether on their own or as parts of larger organisms, are the basic units of life, evolution and selection.

Even if these three cases (classification, evolution and ecology) for the importance of microbes are accepted, people who take a plant or animal perspective might argue that a microbial perspective is all very well for background categories, ancient events and contextual interactions, but that the most interesting biology is nevertheless happening at the multicellular level. I accept that very specific aspects of large-organism biology, such as bone structure and petal colour, can be usefully examined in isolation from microbes. But overall, a fuller understanding of such phenomena and how they came to be the way they are will need to be microbiological. To push this point home, it is worth taking a standard multicellular topic – brain function – to show how this not only has a microbial analogue (see above), but also that to grasp this very activity in multicellular organisms, whether empirically or theoretically, can be greatly illuminated by microbiological insights.

Although it is commonly thought that microbes are not implicated in any way in nervous systems and brains (except as occasional pathogens), this is not so straightforwardly the case. Numerous studies now show that gut microbes have effects on animal brain function, with consequences for mood, cognition, memory and behaviour (Cryan and Dinan 2012; Figure 6.3). There are many environmental influences on brain function, but the microbial pathways are particularly intimate, occurring as part of normal animal development. The gut microbiota has direct and indirect effects on central and peripheral nervous systems via the neural, endocrine and immune systems, with the vagus nerve being a particularly important connection (Heijtz et al. 2011; Figure 6.3). Stressful experiences, for example, affect the gut microbiota, which then feed back into the nervous system. These effects can last for years in adult animals if they occur at

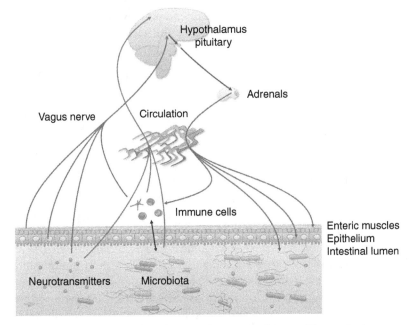

Figure 6.3: Feedback loops between the gut microbiota and human brain.
Based on Cryan and Dinan (2012).

critical developmental periods. Another example is the way in which the presence of unusual pathogens in the gut can trigger behavioural changes in animals (Bercik et al. 2012). More detailed mechanistic study is needed to make all these connections clear. However, studying these cause-effect relationships is given another layer of complication with the requirement of mouse models as experimental systems in which to do human-oriented microbiota and nervous system/brain experiments. Research on how microbes affect brain function in humans is thus inextricably 'multilevel' in the sense used by Love and Travisano (2013).

If even cognition, human behaviour and associated phenomena can be viewed via microbiology, this helps support the claim that starting with microbes is always a good idea for philosophy of biology (and can even, as Dretske and Millikan suggested, supplement philosophy of mind). One problem with doing so is that full agreement with a 'microbes first' position would be very disruptive of how philosophy of biology is practised. It would require refocusing many traditional lines of inquiry. However, the many enhancements provided by a microbiological perspective would compensate for any revisionary work that needs to be done to make philosophy of biology achieve a more inclusive orientation to life science research. But what about philosophy of science? As this chapter has argued, much contemporary practice in the life sciences depends heavily on microbial tools and model systems, and the methods devised in reliance on these microbial models. From phylogenetic reconstruction to experimental evolution, from molecular biology to biochemistry, and from cell biology to large-organism physiology, many of the tools and methods making major advances in today's life sciences have a microbial component. These are important not just to philosophy of biology, but also to philosophy of science in general because of how microorganisms are used as models in every important sense (Love and Travisano 2013; Levy and Currie 2014).

All these considerations add up to an increased sense of the importance of microbes. But a bigger conclusion than this is a necessary outcome of the discussion above. I will suggest in the concluding chapter that micro-biological reflection reorients philosophical understanding of what being human is, and that it displaces humans from the centre of philosophical discussion about anything biological. Understandings of human nature and human interaction – including communication and human cognitive processes – all benefit from and are subtly or sometimes grossly trans-formed in light of microbiology. Microbiological insights thus enhance philosophical reflection in a general sense, by making us think more about the very nature of life itself.

Conclusion: further philosophical questions

This chapter will end the book by addressing some broad questions that arise from the more detailed discussions in previous chapters. These questions have to do with life: how microbiology puts human life into a broader biological context, what life is, and how the study of life could and should be understood by philosophy of biology.

Putting human life into a microbial context

The previous chapters have emphasized a 'continuity of life' perspective that sees all living things as sharing many common capacities. Living entities can be examined within shared scientific frameworks for classification, evolution and ecology, even as this common treatment enables an understanding of major differences between diverse forms of life. There is no need in this loose way of thinking about life to invoke anything such as vitalism, but there is a requirement to understand the physical processes, geology and chemistry that contribute to the origin and evolution of living things. However, despite this unobjectionable continuity view, there is still a tendency amongst philosophers and scientists to perceive humans as special – as something biologically, socially and culturally more extraordinary than the rest of life.

Rather than arguing directly against that view, let me suggest for the final time that a microbiological view puts humans into the right perspective, at least for broad-scope pictures of ecology and evolution. Phylogenetic representations of life's history do this very nicely. As we are all aware, humans are located on a tiny and no doubt fragile twig of any tree of life attempting to represent the entirety of evolutionary history (Figure C.1; also Figure 2.5 in Chapter 2). Moreover, we gain a different picture of evolutionary transitions when microbial metabolism is made the game-changing phenomenon in the history of life. Ecologically, while human impact on the planet is indeed enormous and perhaps disastrous for many

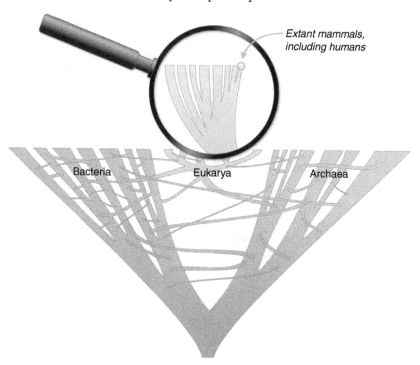

Figure C.1: An impressionistic tree of life, putting humans in phylogenetic context.

large life forms, humans are only tiny nodes in the vast webs of organismal and geochemical relationships sustaining life. And even on an individual-organism view that picks out single humans from those interactions, each of us is an entire ecosystem of microbial activity as the microbiome research in Chapters 3 and 5 showed very clearly. To preserve any sense of human exceptionalism requires lifting ourselves out of our basic biology.

However, there are possibilities of preserving an anthropocentric perspective in some spheres. One of these involves microbial conservation. As discussed in Chapter 5, microbial ecology and specifically biogeography began with the idea that 'everything is everywhere', meaning that there were no geographical constraints on microbial distribution. Now, patterns of distribution that appear similar to those of animals and plants have been detected. These findings lead to questions about conservation, and whether a better understanding of microbial biodiversity and distribution means that a philosophy of microbial conservation and even conservation ethics follow closely behind. Imagine, as seems most likely to be the case, that at least some microorganisms somewhere have very restricted distributions. This would make them rare and even endangered if, for example, their habitats were under threat for any reason. Threats could be factors such as the removal of toxic waste areas (for example, tar ponds or chemical dumps) in which very specialized microbes make a living. With the physical (geographical) and nutritional (environmental) bases of their distributions gone, these microbes could become extinct. Another scenario is that of the smallpox virus. If all potential hosts (non-immune humans) are removed forever – as vaccine campaigns are meant to have achieved in 1989 – the only remnants of this lineage might be deep-frozen viral 'zoos' of isolates. Even these isolates are being targeted for eradication. We usually feel it is tragic when extinction happens to orchids, birds or frogs, but such sentiments are rarely in evidence for microbes.

Although it is still fairly unusual for microbes to be discussed in regard to conservation policy, environmental ethics or anything associated with them, there is some interesting advocacy happening in the microbiological community. Foremost amongst these discussions are essays by geomicrobiologist Charles Cockell (e.g., 2005; 2008), whose work acknowledges earlier seminal papers on microbial conservation by microbiologists Rita Colwell (1997) and Bernard Dixon (1976).[1] Dixon deliberated on the eradication of the smallpox virus. Although he ultimately argued for viral conservation on the grounds of being able to fight newly evolved strains better if old ones were 'conserved', he also suggested that 'if we experience twinges of guilt about the impending extinction of large creatures, why should we feel differently about small ones? Conservationists lavish just as much time and energy on butterflies as they do on elephants. Why discount the microbes?' (Dixon 1976: 432). Nor did he think a case could be made against the conservation of pathogens: 'Every one of the arguments adduced by conservationists applies to the world of vermin and pathogenic microbes just as they apply to whales, gentians and flamingos. Even the tiniest and most virulent virus qualifies' (Dixon 1976: 431).

Taking a broad, biogeochemical perspective, Colwell (1997) focused on the non-viability of life on Earth without microbes. In the same year another well-known microbiologist, James Staley (1997), raised the question of whether microbial species are threatened. Rather than deciding one way or another, he suggested what research would be needed in microbial ecology and taxonomy simply to begin to answer the question. Despite the

---

[1] Cockell (2005) also draws attention to even earlier microbiological conservation arguments (or hints), and to literary discussions of topics such as microbial rights.

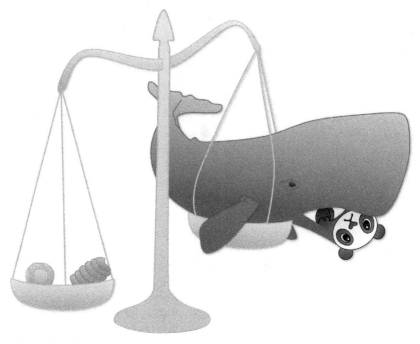

Figure C.2: The balance of conservation: whales and pandas versus ammonia oxidizers.

many gaps in microbiological knowledge, so much more is known now, fifteen years after Staley's manifesto, that it is worth trying to make further progress with this question.

The first step towards conservation microbiology is working out whether lineages could go extinct at least in principle, and then, if they were considered worth preserving, whether anything could be done to prevent such extinctions. But if microbes do not have geographically restricted distributions, this would mean it would be difficult, if not impossible, for them to disappear. And if microbial communities exhibit considerable functional redundancy – as discussed in Chapter 5, in which taxa may vary but the function persists – then the loss of any taxon might not matter to the broader community system, which is over-endowed with the relevant functional pathways. Functions will be preserved by other taxa taking up those same functional roles, and life at the ecosystem or biogeochemical scale will remain more or less the same (Green et al. 2008). Here we see an instance of why the choice of biodiversity unit matters beyond its utility for strictly scientific analysis.

But if some microorganismal lineages indeed could be extinguished, should efforts be made to save them?

> I make no apologies for putting microorganisms on a pedestal above all other living things. For if the last blue whale choked to death on the last panda, it would be disastrous but not the end of the world. But if we accidentally poisoned the last two species of ammonia oxidizers, that would be another matter. It could be happening now and we wouldn't even know (Curtis 2006: 488; Figure 6.5).

Although it is unclear 'we' would even be around to watch the extinction of any microorganisms that maintain the biosphere, this quote captures very nicely the tensions in conservation biology: who or what is conservation for? Us or other biological entities? Nature in all its biodiversity for its own sake?

This question of the purpose of conservation taps into deep issues in environmental ethics, and is usually cashed out in debates about intrinsic versus instrumental value (for a very helpful overview, see Justus et al. 2009). The most practical argument for microbial conservation involves the 'instrumental' conservation of microbes. As Colwell (1997) argued, conservation of the visible life we know and love, including ourselves, will require the conservation of microorganisms. Animals, fungi and plants will not survive without their microorganismal collaborators. Moreover, remediation of many environmental problems such as oil spills will involve microbial interventions, either with the assistance of natural communities, or with communities engineered as bioremediation tools (Atlas and Hazen 2011). Geoengineering the planet out of climate change, if such interventions ever go ahead, is likely to involve major manipulations of microbial communities (e.g., Williamson et al. 2012). And more abstract than all these perfectly good reasons for the human need of microbes is the knowledge argument: 'If many microbial species are lost, there is one intangible benefit that would be lost forever: the knowledge of who we are and what we came from' (Colwell 1997: 305).

But calculating the instrumental value of microbes will take a lot of thought about which units of biodiversity are the units of conservation. Very seldom will individual microorganisms be instrumentally valuable. Populations are probably the least unit that can be conserved, and – as all the discussion in the preceding chapters indicates – communities and ecosystems are the loci of most of the important interactions humans would want maintained (Cockell 2005). Since this value manifests in regard to function rather than composition, a conservation perspective provides another endorsement of microbial ecology's focus on community-level properties. However, if most instrumental value is located in widespread communities, massive microbial extinctions will probably have been preceded by human and other animal extinctions, along with those of most other large organisms for that matter. Conserving microbes will be irrelevant at that point since nobody will be around to calculate instrumental value.

The other side of the value coin involves the intrinsic value of microbial life. Some people might call this the 'deep ecology' conservation of microbes, in which they are conserved simply because they exist. Intrinsic value is harder to make a case for, even for macroorganisms (Justus et al. 2009). The criteria for such attributions are often vague or inconsistent. Cockell (2005) argues that the intrinsic value of microbial life is a position that should be considered seriously, on the basis of microbial life having similar 'interests' to those that ground intrinsic value claims about human life. Microorganisms are alive; they certainly can be said to have interests of their own in surviving, although it is unlikely those interests are accompanied by the same cognitive and emotional perceptions that humans and perhaps other animals have. But even if we were to agree with the notion of microbial interests, it is very hard to respect microbes as intrinsically valuable from a human point of view, especially in the sense of comprehensively refusing to harm them (Cockell 2005). Many everyday human activities destroy microbes and cannot avoid doing so.

It is most probable that an intrinsic value perspective on microbes could be upheld only in the broad sense of awareness of and respect for nature in all its manifestations. This would be an environmental ethic that has 'at its basis the ecological interconnectedness of life [and so] cannot exclude microbes from its purview' (Cockell 2005: 387). Cockell thus allows for intrinsic value as a 'principle' but accepts that a default to instrumental value is what will rule any debate and policy. Although this is essentially a refutation of the intrinsic value position for microorganisms, instrumental value is probably as far as microbial conservation can be taken without being dismissed out of hand. For example, microbiologist Julian Davies (2010) finds the suggestion that microbes could have 'rights' based on intrinsic value 'ridiculous'. Nevertheless, he agrees that any conservation agenda should include microbes, even though microbes look after their own future and can cope with decimation or worse from phage on a daily basis (2010: 721). Compared with these phage-driven mortalities, Davies argues, human interventions are hardly worth worrying about.

But even if dealing with the conservation of microbes on a strictly instrumental basis, how on Earth would endangered lineages of microbes be preserved? There are still enormous numbers of undiscovered and possibly rare microbes whose futures may be in jeopardy due to lack of any knowledge about them. Even if endangered organisms are identified, everything we have discussed so far suggests that understanding organisms in total isolation is not understanding very much about them at all. If isolated cultures frozen in freezers end up being the only way to preserve microorganisms, would this sort of preservation be meaningful in any sense? Perhaps, like Californian condors and red kites, these preserved microorganisms could be reintroduced to the wild in carefully selected and monitored environments. The very unlikeliness of this happening means that in the microbial case, the only really imaginable route to conservation is via general habitat preservation, accompanied by the expectation that the relevant microbes will co-exist with those habitats. These habitats include large charismatic organisms, too. Saving pandas or whales, for example, is likely to preserve the unique microbial variants that are within, on and around these organismal habitats.

Almost all the microbiologists who raise issues of microbial conservation acknowledge that microbes will never be regarded in the same light as pandas (e.g., Staley 1997), and that microorganismal value will always be subordinated to that of larger organisms. Nevertheless, they see room for some concern about microbial conservation in the public eye, and some practical steps that could be taken (Cockell and Jones 2009). And a few conservation policy directives, particularly those associated with habitat monitoring in environments such as soil, have recently acknowledged microbes positively (Bodelier 2011).[2] Yellowstone National Park, discussed in Chapter 5 regarding Brock's discoveries of thermophiles, has a 'microbial diversity conservation' policy and plan, which – alongside the biotechnological exploitation of microbes – aims to preserve habitat and add to culture collections (Varley and Scott 1998). Microbial ecology, and its enormous contributions to knowledge about microbial-environmental interactions, is driving these policy implementations and making it harder to ignore microbes in the science, policy and practical preservation of environments. And microorganisms, for environmental ethicists doing thought experiments about conservation policy, are a rich source of insight that at the very least need to be taken into account, and on many occasions might substitute very effectively for pandas and whales.

---

[2] However, see Cockell (2005) for an example of the explicit exclusion of microbes in conservation policy.

The questions these reflections on microbial conservation lead to, however, are difficult ones. Does a microbiologically contextualized view of life say anything about human responsibilities and obligations? Are we just the passive recipients of microbial largesse, or is some other relationship implied? Given how difficult it is to change social and economic practices even in light of extinctions of charismatic organisms and the destruction of visible habitats, it might seem that a microbiological view of life won't make any difference. If concern for big life forms has not been enough to alter human behaviour, it would be surprising if awareness of or even concern about microbes changed anything. But perhaps by taking a more closely connected view of living things, and possibly by seeing how even the smallest entities contribute to human well-being, life as we know and prefer it can be sustained. Realistically, however, the planet's future is not ours but belongs to our microbial collaborators.

The nature of life

But given the past and present, microbes and microbiology tell philosophers and scientists a great deal about living things. 'What is life?' is a question with a long pedigree in natural philosophy. Aristotle (384–322 BCE) and Immanuel Kant (1724–1804) are the philosophers most invoked in a philosophical tradition of reflecting on the nature of living things (Grene and Depew 2004), but the history of microbiology itself repeatedly throws up philosophical quandaries about how life arises from matter and whether it can change once it has arisen. The former topic is the crucial issue in debates over spontaneous generation, which gained their initial momentum from the microscopy of the 1600s and were subjected to increasingly rigorous experimentation in the eighteenth and nineteenth centuries (Farley 1974). Although claims and counterclaims about spontaneous generation continued into Darwin's era, a different issue was foregrounded by increasing acceptance of evolution – not necessarily evolution by natural selection, however. This debate was about the capacity of one form of life to transform evolutionarily into a new one, and was very keenly disputed in nineteenth and early twentieth-century microbial classification (see Chapter 3). The evidence for and against species transformations in every realm of life was marshalled under deeply philosophical commitments (Sloan 2008), and contemporary biology's search for a universal species concept might be seen as a residue of this older debate. But despite a growing consensus in twentieth-century biology that all organisms could be classified within hierarchies that reflected evolutionary processes, the overarching question remained about how the line between non-living and living had been crossed. Spontaneous generation experimentation turned into origin-of-life research, and molecular biology eventually took centre stage in those efforts.

As molecular biology emerged in the 1940s, it was accompanied by physicist Erwin Schrödinger's thoughts on 'What is Life?' (1944). Although the extent to which Schrödinger influenced molecular biology is debatable (Yoxen 1979), what he did do, at least in retrospect, was to reaffirm in theoretical terms that questions about life had to focus on mechanisms of heredity and metabolism. Although Schrödinger is most known to philosophers for theorizing about an orderly 'aperiodic crystal', which the discovery of DNA structure appeared to confirm (Kay 2000), he was at least as interested in metabolism and its contribution to answering his question. Schrödinger did this from a thermodynamic perspective that was not concretely biochemical (Weber 2011), but he nonetheless captured a theoretical tension from earlier work that was still unresolved.

Origin-of-life models had already emerged in line with this sort of division, with the experimental and theoretical pendulum swinging between metabolism-first and replication-first accounts. Respectively, they award explanatory primacy in the origins of life either to a basic process of autocatalysis or to some form of nucleic acid.[3] Almost all these models nowadays require some sort of compartmentalization: naked nucleic acids or uncontained chemical interactions are not thought to be the starting point of life (Fry 2010). Although any origin-of-life theory is going to agree that microbial life forms are the next step from a protocellular world, it is viruses I will focus on here, primarily for their philosophical interest. Viruses are an excellent model system for philosophical examination of theories of life and origins-of-life, even though viral entities are not usually considered to be alive. They do replicate, but only with the assistance of cellular entities, and they do not metabolize – having neither the necessary genes nor the cellular capacities with which to do so.

Virus research emerged in the late nineteenth century, and scientific accounts of the evolution of viruses have changed quite radically over the twentieth century. First thought of as precursors to cells, viruses are now largely considered to be selfish cellular escapees – bits of a genome that replicate by taking advantage of genuine reproducers, which are necessarily cellular (Holmes 2011). However, this standard evolutionary narrative has

---

[3] There are also very abstract theories of biological organization that have some bearing on origin-of-life models. These theories are summarized in Letelier et al. (2011).

been and continues to be challenged, with a few current models proposing that viruses are indeed cellular precursors (e.g., Koonin et al. 2006). Viruses are sometimes argued not only to be as ancient as cells, but also to play causal roles in the formation of the cell types now known as bacterial, archaeal and eukaryotic (Forterre 2006; Claverie 2006).

Although these different evolutionary scenarios cannot yet be arbitrated with current data and methods, and may never be – because the right kind of data may never be available, and because affinities for particular hypotheses run deep – there is a second parallel discussion about viruses that is very informative for philosophical discussions of the nature of life. Recent microbiological findings of extremely large viruses have inspired arguments that some viruses fall between cellular and acellular life forms (see Chapter 2). In one line of interpretation, the traditional association of life with cellularity is considered a mistake. Instead of making a division between cellular and non-cellular biological entities, say microbiologists Didier Raoult and Patrick Forterre, the biological world should be divided into 'ribosome-encoding organisms' and 'capsid-encoding organisms', which when reproducing should be understood as 'ribocells' and 'ribovirocells' (2008: 317; Forterre 2013). Capsids are the protein coats encapsulating viruses. They come in many different morphologies, and in some viruses may also be enveloped by membranes. Emphasizing capsids, which are unique to viruses, is done as a corrective to a cell-based focus on life.

Despite the proliferation of novel virus insights due to the development of new molecular and visualization tools, debates about the status of these entities are heating up rather than cooling down. For this reason, viruses should be of great interest to philosophers, who – even if not interested in biological minutiae – surely should be concerned with these very basic philosophical questions about what life is and when it began. There is no question that microbes (including viruses) can help inform any such philosophical inquiry. Whether any robust consensus on a definition of life can ever be found is currently an open question (Cleland and Chyba 2002). But even with hypothesized gains in knowledge about extra-terrestrial biology, and experimental advances in establishing origin-of-life scenarios, philosophy will continue to have a role: it is unlikely that philosophical discussion of life exists just because there are scientific gaps. The persistence of philosophy in the discussion is for the worse, says biochemist and geneticist Jack Szostak (2012), who claims that philosophical efforts to define life are futile, irrelevant and perhaps even harmful to the scientific project of understanding its origins. However, given the unlikeliness of definitive and complete answers about life being provided from analyses of the evolutionary past or future experimental achievements, it makes sense to recognize the philosophical nature of questions about the properties of life, and for general reflections on concepts and theories of life to aim to enable further empirical and theoretical inquiry.

In addition to philosophical interrogations of microbiology, in which microbiology provides a framework for the questions that could be asked, there could be a constructive philosophy-led project, too. A candidate for this is a general account of life (not a definition) that might be indicated by the microbial world. What does all the interconnectedness and collaborative association discussed in the preceding chapters mean for how we understand being alive? One of the reasons why viruses are rejected from the category of living things is that they display very limited autonomy. However, a major message of the previous chapters has been that the very entities we have so often thought of as free-living, independent and self-sustaining organisms in fact are entangled in deep webs of interdependence, both evolutionarily and ecologically. Philosophers and scientists who hold methodological individualism as the most reasonable way in which to see the world can recognize those connections, but then argue that bigger biological units are always formed from the most basic unit of life: the ecologically and evolutionarily isolated individual organism. If, however, a more collectivist perspective is taken, the primacy is reversed: the collectives become the basic unit of life and the individual organisms the by-product of larger organizational strategies. This does not entirely do away with hierarchy in the sense of nestedness, but it does suggest that strict notions of biological hierarchy are open to challenge (see below).

Evolutionary transitions also need reconceptualizing when considered from this collectivist microbiological point of view. I have already discussed one possibility in Chapter 1, where an alternative 'metabolism-focused' model of evolutionary transitions was proposed. Let me return to this topic briefly to bring out even more generally what this alternative implies. In an illuminating essay in 1999, Sterelny compared Gould's model of evolutionary transitions with Szathmáry and Maynard Smith's scheme (Sterelny 1999). Gould's position, presented discursively in *Full House* (Gould 1997),[4] is a 'bacterial' (that is, prokaryotic) view of evolution. In Sterelny's summation of Gould's view, 'life began at, or soon reached,

---

[4] *Full House* was published in the UK in 1997 with the main title *Life's Grandeur*, which is the edition cited in the reference list.

a bacterial mode: most living things are bacteria. It has stayed that way. Almost every living thing that lives now, or ever has lived is a bacterium' (Sterelny 1999: 463). Gould does not emphasize 'major' transitions and directionality in the way Maynard Smith and Szathmáry do, because everything occurs against this backdrop of prokaryotic diversification and persistence.

Although Sterelny prefers the Maynard Smith and Szathmáry approach due to its provision of a more directional model of a variety of changes in biological organization, he sees the two views as compatible. But at the same time he recognizes that taking a metabolic rather than a morphological focus is unlikely to support an 'increasing complexity' hypothesis (Sterelny 1999: 466). What would be lost in a microbial metabolism-focused view of evolutionary transitions is the notion of increasing hierarchy, says Sterelny: 'vertical complexity [as] ... the number of layers or nestings within a system ... cell, tissue, organ, organ system, organism' (Sterelny 1999: 467). Theorizing hierarchy is very important to philosophers, whether for evolutionary or organizational discussion, but it is also problematic because of the many different dimensions of organization that hierarchical schemes try to capture (Jagers op Akkerhuis 2008; Love 2012; Potochnik and McGill 2012). Although it may seem intuitive to recognize the supposedly natural levels in a traditional biological hierarchy, the metabolism-focused view moves the emphasis from these 'levels' to what keeps these groupings together over physiological and evolutionary time. Metabolism is a good candidate for this purpose because it explains a great deal about evolving entities such as cells and cellular collectives. It provides answers to questions about why units of adaptation do not map one-to-one onto genetically isolated lineages, which are the foci of the standard approach to evolutionary transitions. With a metabolic focus, evolutionary transitions can be viewed in terms of 'horizontal complexity' rather than the vertical complexity captured in a directional model.

Flattening an entrenched hierarchical view of life does not mean, however, rejecting the usefulness of thinking at appropriate moments about particular structures of nested systems (e.g., Okasha 2011; Haber 2013). But the loss of any absolute hierarchy does imply that nestedness is not the only way to think about biological organization, and that much may be missed – physiologically, ecologically, and evolutionarily – when it is the overarching model of organization. An interactive metabolism-focused view of life does not require the imposition of a great deal of replacement conceptual machinery: examination of metabolic relationships over time is enough to get started. This also gives considerable insight into major questions arising from contemporary molecular microbiology about biological and Darwinian individuals. When metabolism is emphasized, these individuals tend to be more inclusively conceptualized based on functional collaboration, and are not specified primarily by genetic lineages. Taking this perspective is helpful at all phases of evolution, from the precellular world and into the cellular world, with all the transformations of metabolic capacities that cellularity enables. It is also less vulnerable to the chauvinism of a directional view, which in Maynard Smith and Szathmáry's model culminated in human communicative abilities. However, doing philosophy of biology from this 'flatter' perspective naturally raises questions about the very approach to understanding life.

How the study of life can be understood by philosophy

From a philosophy of biology point of view, there are at least two good reasons to pay attention to microbiology. First, it is useful to compare answers to traditional philosophical questions asked of other biological fields with the answers when those same questions are asked of microbiology. Second, there may be new philosophical questions that arise in regard to microbiology but which are not generated in traditional fields of scrutiny, such as evolutionary biology carried out within a zoological perspective. The preceding chapters deal with both types of questions in the discussions of microbial classificatory practice, evolution and ecology. Each chapter showed how traditional questions – about, for instance, what a species or a unit of selection is – were put in a larger biological context by including microbes, and the original questions at least slightly transformed in the process. In addition, new questions could be asked about phenomena such as reproduction, for example, when lineages and functional persistence do not always coincide.

But to take an even broader overview it will be helpful to reflect on how philosophers of biology have approached a large field of practice such as molecular biology. As noted in Chapter 6, microbes were and still are immensely important models for molecular studies. Although philosophy of biology has been greatly interested in molecular biology, it has focused fairly abstractly on the field, most commonly via questions about information, reduction, determinism and gene conceptualization. In that scrutiny, the model systems largely dropped out of the philosophical picture. As the microbial bases vanished, so did another molecular process of crucial importance to molecular analysis: the biochemistry of metabolism. Biochemistry, although it has been a major field for over a century, has not been a popular object of study for philosophers,[5] despite the way it is woven into early molecular biology, and despite its crucial importance for understanding organismal function and evolution.

Philosophers of science Lindley Darden and James Tabery illustrate this position very nicely when they suggest that historically,

> biochemistry was concerned with nutrition (recharacterized as metabolism more generally) and molecular biology (along with its more direct predecessor classical genetics) investigated reproduction ... Much of biochemistry's focus *(for the perspective of what is important for molecular biology's questions about the genetic material)* was on proteins and enzymes (Darden and Tabery 2009: unpaginated, emphasis added).

Molecular biologists, as they effortfully distinguished themselves from biochemists, relegated biochemistry to a support role for the synthesis of nucleic and amino acids (Davis 2003). Acceptance of this status allowed philosophy of biology to focus on genetic molecules and morphology, with the latter usually anatomy rather than the morphology of intracellular organelles. Metabolism, which could be thought of as the connective in the middle, has been passed over.

I discussed above and in previous chapters how a focus on metabolism might change philosophical perspectives on evolutionary individuals and transitions, and the same perspective also would change the general organismal focus in philosophy of biology. If metabolism had a higher epistemic profile, then microbes would be philosophically important biological systems rather than just those with legs, hearts and nervous systems. And biochemistry itself is a fast-moving science that captures dynamic complex processes from both theoretical and experimental perspectives. Its models and methods are sophisticated, contested and still evolving (e.g., Chen et al. 2010). Philosophy of microbiology would encourage more attention to biochemistry, and thus fill another major gap in philosophical attention.

But for many philosophers of biology, emphasizing biochemistry would ultimately lead to the question of whether metabolism or genetics is the most important explanatory focus. Another way to think about this is in regard to the questions raised by pluralism: is philosophy of microbiology suggesting a plurality of views, or is taking a microbiological perspective unificationist in some sense? We could think about microbiology and

---

[5] There are of course exceptions to this generalization. These are mostly philosophical discussions of the origins and definitions of life, as well as some specific episodes in the history of biochemistry. However, they do not overwhelm a more standard philosophical focus on genetic material and processes.

pluralism in two ways. One is that microbiology gives a foundational understanding of life, onto which the quirks of macrobial biology are simply added. This position is along the lines of what philosopher Carol Cleland (2013) is suggesting. The other is that microbiology discloses an overwhelming variety of living things and processes, such that no unified account will ever be possible. This is probably exemplified most strongly in philosopher John Dupré's position on pluralism (1993), even though it was not originally addressed to microbes and microbiology (it is now). A less metaphysical but still strong justification of pluralism (but without microbes) is 'empirically motivated', and based on observations of scientific practice in which multiple accounts appear to be needed to make sense of the world (Kellert et al. 2006; Love 2012).

Although these are important perspectives in philosophy of science, it is unlikely the options are as simple as a dichotomous forced-choice, however. For many scientists, 'pluralism' is a good thing, to be used in the service of a unified understanding and the best explanation of any phenomenon. In this sense, scientists are often pragmatic unificationists, attempting to find the best explanation, but also methodological pluralists, willing to use as many methods as it takes to reach that unified or single-best explanation. Being an explanatory pluralist, in the sense of holding that incompatible explanations are equally good for whatever the explanandum is, is a risky epistemic route to take. It leads to the problem of pressing on with one's own research agenda and potentially not worrying enough about what other scientists are finding and how they explain it. Accepting a plurality of incompatible explanations in advance of inquiry would not produce the kind of science we think exists today, where there is competition between explanations to be accepted as the best. So explanatory pluralism is not *prescriptive* in this sense, although it may be in a much weaker sense of different modes of explanation (for example, statistical and causal) or different models being seen as applicable to the same target phenomenon (Brigandt 2012; Weisberg 2007). Instead, pluralism is more likely to be a *diagnosis* of a situation where there are strong reasons for not subordinating or replacing one explanation or conceptual apparatus with a competing account. Species concepts in their present state fit this situation, and that fit reinforces the problematization of hierarchy. Normative methodological pluralism, however, is something fairly uncontentious in comparison: very rarely would it be said that tackling a research question with a variety of techniques, methods and models would be undesirable. This conclusion about methodological pluralism can be applied more broadly to the relationship between philosophy and science.

There are several points of view about how philosophy of biology and science do or should interact with science. Some commentators have seen this relationship as corrective or at least deflationary – in which philosophy deflates or corrects the science – while others have perceived a master-builder (science) and underlabourer (philosophy) relationship. From a practical rather than a meta-methodological angle philosopher of palaeontology Derek Turner distinguishes two approaches to philosophy of science. The first is a top-down or 'philosophy first' approach. It starts with standard philosophical questions about laws, causation, reduction, the nature of evidence and realism. The second, which he calls 'bottom-up' or 'science-first' philosophy, involves starting with the science and working upwards into the philosophical implications.

> Philosophers who go this route usually try to learn as much as they can about developments on the ground in the field they want to study, and they try to get wrapped up in some of the same questions that occupy the scientists. However, bottom-up philosophy of science, when done well, is not merely descriptive. The challenge is to begin with the science and then gradually work one's way up into philosophical territory (Turner 2011: 12).

In Chapter 1, I showed how taking a strong 'philosophy-first' approach in teleosemantics could develop conceptual machinery for philosophers but not have any feedback into the science itself. At the other extreme, of science-first philosophy, there is the rather different problem of how to do philosophy with science such that there is more than just reiteration of what is being done and talked about perfectly well by the scientists themselves.

Practical methodological concerns such as these raise questions about the boundaries between science and philosophy and what lies between them. In Turner's words,

> In bottom-up philosophy of science, where does the discussion of science end and the discussion of philosophy begin? Sometimes it may be easy to tell when one has crossed the frontier from natural science into philosophy, from the empirical into the world of the conceptual and normative. Sometimes, though, the frontier is vague and difficult to make out; questions of theoretical definition ... interweave with questions about the world ... in a way that often makes it difficult to tell whether one is talking about ideas or about things in the world. Natural science and philosophy are like two countries on a map with a common border that is well marked in some places, disputed in others, and in still other places completely undefined. It's possible to start in the middle of one country and head in the general direction of the other one without being able to say, or even much caring, when one has crossed the border (Turner 2011: 14).

In this book, I have rather obviously tried to follow what Turner calls the 'bottom-up' approach, and I think microbiology very much lends itself to doing philosophy of biology and science via this route. This is in part because microbiology is new territory for most philosophers, and using standard philosophical tropes and classic questions might artificially constrain analyses of this philosophically uncharted science. And the other reason for doing it this way – pushing from the scientific end – is that the old questions and conundrums of philosophy are usefully affirmed if new forms of inquiry end up there anyway.

In this respect, I have taken a cautious route in this book, directing the microbiological topics I have examined toward familiar philosophical themes. Microbiology clearly has a great deal to say to well-worn issues in philosophy of biology about natural kinds and units of selection; it also has important things to say to newer, less jaded topics about modelling and multiscale interactions. And microbiology may also encourage a different way of looking at biological organization, so that philosophers foreground metabolic collaboration and put in the background hierarchies of ever-more inclusive systems. Part of the reason for doing philosophy of microbiology with existing philosophical questions, rather than arguing there are totally new philosophical issues to be considered, is strategic. Because the neglect of microbiology in philosophy of biology still needs to be remedied, making connections with familiar topics – which have sometimes been only half thought-through because microbes have been left out – is likely to persuade more philosophers that microbes matter. Methodologically, therefore, philosophy of microbiology has an interesting middle road to tread, as it wends its way between showing its relevance to philosophy of science in general and philosophy of biology in particular, and revealing to scientific communities the important philosophical aspects of even the most contemporary developments in microbiological science. Naturally, there are limits to striking out from the science and hoping to end up somewhere useful, and there is always going to be a compromise between bottom-up and top-down philosophical approaches. This compromise reflects the fuzzy boundaries between science and philosophy and focuses on their overlaps.

Final reflections

The main message of all the preceding chapters is that when philosophers are thinking about conceptual machinery and modelling strategies in biology, and big topics such as classification, evolution, ecology and life, then starting with microbes and microbiological research is a highly effective tactic. Whether microbes become a supplementary topic or a major area of investigation for the philosophy of biology is not the main issue: a more adequate understanding of life simply requires that they be recognized in many different contexts. A stronger formulation of this view is that philosophers really need to start making their biology-based models with microbes first, before theorizing more specifically about their favourite animals. But even the weaker view – 'keep microbes in mind' – will bring about many changes in philosophy of biology. As this accommodation occurs, some microbiologists and other biologists might be persuaded that their fields have deep and abiding philosophical issues that guide investigation and sometimes obstruct it. Whatever that embedded philosophy is doing, it is worth being aware of, and philosophy of microbiology can contribute to that awareness.

Glossary

Metabolism and energy generation

Chemoautotrophy:
The ability to use inorganic rather than organic carbon sources, but with inorganic substrates rather than light as an energy source (as opposed to photoautotrophy).

Chemoorganotrophy (or chemoheterotrophy):
Metabolisms that use organic compounds for their carbon sources.

Mixotrophy:
Metabolisms that combine autotrophy with heterotrophy. Strategies are different in protists and prokaryotes but any mixotroph can switch from being a primary producer to being a consumer (not necessarily by preying directly on other organisms), and back again.

Plastid:
The organelle (a membrane-bound processing unit in a cell) in photosynthesizing eukaryotes, or in previously photosynthesizing eukaryotes, since function can be lost. Plastids are the genomically and physiologically reduced version of the ancestral cyanobacterial endosymbiont.

Chloroplast:
The label given to the plastid in green plants and some algae.

Systematics, species and evolution

Exaptation:
: This neologism for 'preadaptation' refers to characteristics that may have been adaptive in regard to an earlier function. That function then changed and was selected for in a new role. Feathers are the standard example: they were initially selected for their insulating properties and only subsequently for their contribution to flight.

Monophyletic group:
: Includes the most recent common ancestor and all its descendants (see Figure G.1).

Paraphyletic group:
: Includes the most recent common ancestor but not all the descendants (see Figure G.1).

Polyphyletic group:
: Does not include the most recent common ancestor (see Figure G.1).

Allopatric speciation:
: The bifurcation of a population due to geographical isolation, after which gene flow is prevented and subsequently maintained even if the geographical barriers are removed.

Sympatric speciation:
: The divergence of a population that remains in the same spatial location but in which ecological differences bring about an end to gene flow between sub-populations.

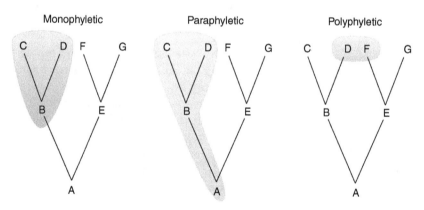

Figure G.1: Monophyletic, paraphyletic and polyphyletic groups.

Particular microbial phenomena

Chemotaxis: A motility mode for cells in which they direct themselves along or against chemical gradients. In prokaryotes, chemotaxis depends on chemical receptors altering the rotation of the flagella. Bacteria and archaea 'run' and 'tumble', depending on signals they pick up from the chemicals. A reversible chemical alteration (methylation) on the chemoreceptors functions as 'memory' of the chemical gradient, causing the organisms to stay in the same place or keep searching.

CRISPR: The prokaryote phage protection system known as CRISPR-Cas (clustered, regularly interspaced palindromic repeats-CRISPR associated proteins) consists of tiny genetic sequence repeats that confer resistance to phages and other foreign DNA invaders. DNA from the foreign entity is integrated into the CRISPR section of the prokaryote genome, activating the CRISPRs, which in transcript form target the phage sequence for destruction. This is considered Lamarckian by some evolutionists because the prokaryote genome is modified in response to the phage, in a way that is adaptive for the prokaryote and transiently heritable. Thomas Pradeu (2012) discusses CRISPRs in his philosophical account of immune systems and biological individuality.

Hypermutation: A constitutive capacity for increased mutation that is possessed by only a subpopulation of 'mutators'. Stress-induced hypermutation is a transient capacity turned on in large numbers of cells by environmental conditions.

Retrovirus: An RNA virus that uses a cell to reverse-transcribe itself to DNA, and then become part of the host cell's genome as a provirus. The provirus replicates either by producing new RNA viruses, or by being copied as the host genome DNA is replicated. When this viral genome is in a germ cell, the virus is called an endogenous retrovirus.

References

Abreu F, Silva KT, Martins JL, Lins U (2006) Cell viability in magnetotactic multicellular prokaryotes. *International Microbiology* 9: 267–272

Achtman M, Wagner M (2008) Microbial diversity and the genetic nature of microbial species. *Nature Reviews Microbiology* 6: 431–440

Ackermann H-W (2011) Bacteriophage taxonomy. *Microbiology Australia* May: 90–94

Ackert LT Jr (2006) The role of microbes in agriculture: Sergei Vinogradskii's discovery and investigation of chemosynthesis, 1880–1910. *Journal of the History of Biology* 39: 373–406

Adl SM, Simpson AGB, Lane CE, Lukeš J, Bass D, Bowser SS, Brown MW, Burki F, Dunthorn M, Hampl V, et al. (2012) The revised classification of eukaryotes. *Journal of Eukaryotic Microbiology* 59: 429–493

Adler J (1966) Chemotaxis in bacteria. *Science* 153: 708–716

—(2011) My life with nature. *Annual Review of Biochemistry* 80: 42–70

Aguilar C, Vlamakis H, Losick R, Kolter R (2007) Thinking about *Bacillus subtilis* as a multicellular organism. *Current Opinion in Microbiology* 10: 638–643

Allen HK, Donato J, Wang HH, Cloud-Hansen KA, Davies J, Handelsman J (2010) Call of the wild: antibiotic resistance genes in natural environments. *Nature Reviews Microbiology* 8: 251–259

Allen JF, Martin W (2007) Out of thin air. *Nature* 445: 610–612

Allison SD, Martiny JBH (2008) Resistance, resilience, and redundancy in microbial communities. *Proceedings of the National Academy of Sciences USA* 105: 11512–11519

Almquist E (1909) Linné und die Mikroorganismen. *Medical Microbiology* 63: 151–176 [Oral translation provided by Mathias Grote, 2010]

Alonso CR, Wilkins AS (2005) The molecular elements that underlie developmental evolution. *Nature Reviews Genetics* 6: 709–715

Amann RI, Ludwig W, Schleifer K-H (1995) Phylogenetic identification and in situ identification of individual microbial cells without cultivation. *Microbiological Reviews* 59: 143–169

Andam CP, Gogarten JP (2011) Biased gene transfer in microbial evolution. *Nature Reviews Microbiology* 9: 543–555

Anderson W (2004) Natural histories of infectious disease: ecological vision in twentieth-century biomedical science. *Osiris* 19: 39–61

Andersson JO (2009) Gene transfer and diversification of microbial eukaryotes. *Annual Review of Microbiology* 63: 177–193

Andrews JH (1991) *Comparative Ecology of Microorganisms and Macroorganisms.* NY: Springer-Verlag

Angert ER (2005) Alternatives to binary fission in bacteria. *Nature Reviews Microbiology* 3: 214–224

—(2012) DNA replication and genomic architecture of very large bacteria. *Annual Review of Microbiology* 66: 197–212

Ankeny RA, Leonelli S (2011) What's so special about model organisms? *Studies in History and Philosophy of Science* 42: 313–323

Appeltans W, Ahyong ST, Anderson G, Angel MV, Artois T, Bailly N, Bamber R, Barber A, Tartsch I, Berta A, et al. (2012) The magnitude of global marine species diversity. *Current Biology* 22: 2189–2202

Archibald JM (2009) The puzzle of plastid evolution. *Current Biology* 19: R81–R88

Archie EA, Theis KR (2011) Animal behaviour meets microbial ecology. *Animal Behaviour* 82: 425–436

Arumugam M, Raes J, Pelletier E, Le Paslier D, Yamada T, Mende DR, Fernandes GR, Tap J, Bruls T, Batto J-M, et al. (2011) Enterotypes of the human gut microbiome. *Nature* 473: 174–180

Atkinson S, Williams P (2009) Quorum sensing and social networking in the microbial world. *Journal of the Royal Society Interface* 6: 959–978

Atlas RM, Hazen TC (2011) Oil degradation and bioremediation: a tale of the two worst spills in U.S. history. *Environmental Science and Technology* 45: 6709–6715

Azam F, Fenchel T, Field JG, Gray JS, Meyer-Reil LA, Thingstad F (1983) The ecological role of water-column microbes in the sea. *Marine Ecology Progress Series* 10: 257–263

Baas Becking LGM (1934) *Geobiologie of Inleiding tot de Milieukunde.* The Hague: van Stockum and Zoon

Bailey J (1997) Building a plasmodium: development in the acellular slime mould *Physarum polycephalum. BioEssays* 19: 985–992

Bailey JV, Salaman V, Rouse GW, Schulz-Vogt HN, Levin LA, Orphan VJ (2011) Dimorphism in methane seep-dwelling ecotypes of the largest known bacteria. *ISME Journal* 5: 1926–1935

Balaban NQ (2011) Persistence: mechanisms for triggering and enhancing phenotypic variability. *Current Opinion in Genetics and Development* 21: 768–775

Balaban NQ, Merrin J, Chait R, Kowalik L, Leibler S (2004) Bacterial persistence as a phenotypic switch. *Science* 305: 1622–1625.

Balázsi G, van Oudenaarden A, Collins JJ (2011) Cellular decision making and biological noise: from microbes to mammals. *Cell* 144: 910–925

Baltimore D (1971) Expression of animal virus genomes. *Bacteriological Reviews* 35: 235–241

Baltrus DA (2013) Exploring the costs of horizontal gene transfer. *Trends in Ecology and Evolution* 28: 489–495

Bamford DH (2003) Do viruses form lineages across different domains of life? *Research in Microbiology* 154: 231–236

Bapteste E, O'Malley MA, Beiko RG, Ereshefsky M, Gogarten JP, Franklin-Hall L, Lapointe FJ, Dupré J, Dagan T, Boucher Y, Martin W (2009) Prokaryotic evolution and the tree of life are two different things. *Biology Direct* 4: 34, doi:10.1186/1745-6150-4-34

Bass D, Richards TA (2011) Three reasons to re-evaluate fungal diversity 'on Earth and in the ocean'. *Fungal Biology Reviews* 25: 159–164

Baumberg S, Young JPW, Wellington EMH, Saunders JR (eds) (1995) *Population Genetics of Bacteria*. Cambridge, UK: Cambridge University Press.

Bayles KW (2007) The biological role of death and lysis in biofilm development. *Nature Reviews Microbiology* 5: 721–726

Bazylinksi DA, Frankel RB (2004) Magnetosome formation in prokaryotes. *Nature Reviews Microbiology* 2: 217–230

Beadle GW, Tatum EL (1941) Genetic control of biochemical reactions in *Neurospora*. *Proceedings of the National Academy of Sciences USA* 27: 499–506

Beatty A, Ehrlich P (2010) The missing link in biodiversity conservation. *Science* 328: 307–308

Beijerinck MW (1900–1901) On different forms of hereditary variation in microbes. *Proceedings of the Koninklijke Nederlandse Akademie van Wetenschappen* 3: 352–365. Available at http://www.digitallibrary.nl

—(1913) De infusies en de ontdekking der bakterien. *Jaarboek van de Koninklijke Akademie van Wetenschappen* 1–28

Bell G (2010) Fluctuating selection: the perpetual renewal of adaptation in variable environments. *Philosophical Transactions of the Royal Society London B* 365: 87–97

Bell T, Ager D, Song J-I, Newman JA, Thompson IP, Lilley AK, van der Gast CJ (2005) Larger islands house more bacterial taxa. *Science* 308: 1884

Ben-Jacob E, Becker I, Shapira Y, Levine H (2004) Bacterial linguistic communication and social intelligence. *Trends in Microbiology* 12: 366–372

Benton MJ (2000) Stems, nodes, crown clades, and rank-free lists: is Linnaeus dead? *Biological Review* 75: 633–648

Bercik P, Collins SM, Verdu EF (2012) Microbes and the gut-brain axis. *Neurogastroenterology and Motility* 24: 405–413

Berg HC (1975) Chemotaxis in bacteria. *Annual Review of Biophysics and Bioengineering* 4: 119–136

Besemer K, Peter H, Logue JB, Langenheder S, Lindström ES, Tranvik LJ, Battin TJ (2012) Unraveling assembly of stream biofilm communities. *ISME Journal* 6: 1459–1468

Beveridge TJ (2001) Use of the Gram stain in microbiology. *Biotechnic & Histochemistry* 76: 111–118

Bivins R (2000) Sex cells: gender and the language of bacterial genetics. *Journal of the History of Biology* 33: 113–139

Blakemore RP (1975) Magnetotactic bacteria. *Science* 190: 377–379

Blakemore RP, Frankel RB (1981) Magnetic navigation in bacteria. *Scientific American* 245: 58–65

Blankenship RE (2010) Early evolution of photosynthesis. *Plant Physiology* 154: 434–438

Blankenship RE, Sadekar S, Raymond J (2007) The evolutionary transition from anoxygenic to oxygenic photosynthesis. In Falkowski PG, Knoll AH (eds), *Evolution of Primary Producers in the Sea* (pp. 21–35). Amsterdam: Elsevier

Blount ZD, Borland CZ, Lenski RE (2008) Historical contingency and the evolution of a key innovation in an experimental population of *Escherichia coli*. *Proceedings of the National Academy of Sciences USA* 105: 7899–7906

Bodelier PLE (2011) Toward understanding, managing, and protecting microbial ecosystems. *Frontiers in Microbiology* 2: 80, doi:10.3389/fmicb.2011.00080

Bodył A, Mackiewicz P, Gagat P (2012) Organelle evolution: *Paulinella* breaks a paradigm. *Current Biology* 22: R304–R306

Bolker JA (2009) Exemplary and surrogate models: two modes of representation in biology. *Perspectives in Biology and Medicine* 52: 485–499

Bonfante P, Genre A (2008) Plants and arbuscular mycorrhizal fungi: an evolutionary-developmental perspective. *Trends in Plant Science* 13: 492–498

Bonfante P, Genre A, Anca I-A (2009) Plants, mycorrhizal fungi, and bacteria: a network of interactions. *Annual Review of Microbiology* 63: 363–383

Borenstein E (2012) Computational systems biology and *in silico* modeling of the human microbiome. *Briefings in Bioinformatics* 13: 769–780

Boschetti C, Carr A, Crisp A, Eyres I, Wang-Koh Y, Lubzens E, Barraclough TG, Micklem G, Tunnacliffe A (2012) Biochemical diversification through foreign gene expression in bdelloid rotifers. *PLoS Genetics* 8(11): e1003035

Botstein D, Maurer R (1982) Genetic approaches to the analysis of microbial development. *Annual Review of Genetics* 16: 61–83

Bouchard F (2008) Causal processes, fitness, and the differential persistence of lineages. *Philosophy of Science* 75: 560–570

Bouchard F, Huneman P (2013) *From Groups to Individuals: Evolution and Emerging Individuality*. Cambridge, MA: MIT Press

Boyer M, Madoui MA, Giminez G, La Scola B, Raoult D (2011) Phylogenetic and phyletic studies of informational genes in genomes highlight existence of a 4th domain of life including giant viruses. *PLoS One* 5(12): e15530, doi:10.1371/journal.pone.0015530

Bracht JR, Fang W, Goldman AD, Dolzhenko E, Stein EM, Landweber LF (2013) Genomes on the edge: programmed genome instability in ciliates. *Cell* 152: 406–416

Breidenmoser T, Engler FO, Jirikowski G, Pohl M, Weiss DG (2010) *Transformation of Scientific Knowledge in Biology: Changes in Our Understanding of the Living Cell Through Microscopic Imaging*. Berlin: Max-Planck-Institut Für Wissenschaftgeschichte

Brigandt I (2012) Explanation in biology: reduction, pluralism, and explanatory aims. *Science and Education* 22: 69–91

Brock TD (ed and transl) (1961) *Milestones in Microbiology*. Englewood Cliffs, NJ: Prentice-Hall

—(1966) *Principles of Microbial Ecology*. Englewood Cliffs, NJ: Prentice-Hall

—(1967) Life at high temperatures. *Science* 158: 1012–1019

—(1987) The study of microorganisms in situ: progress and problems. *Symposium of the Study for General Microbiology* 41: 1–17

—(1990) *The Emergence of Bacterial Genetics*. Cold Spring Harbor, NY: Cold Spring Harbor Press

—(1995) The road to Yellowstone – and beyond. *Annual Review of Microbiology* 49: 1–28

Brookfield JFY (2001) The evolvability enigma. *Current Biology* 11: R106–R108

Brown MW, Kolisko M, Silberman JD, Roger AJ (2012) Aggregative multicellularity evolved independently in the eukaryotic supergroup Rhizaria. *Current Biology* 22: 1123–1127

Brown, JH (1932) The biological approach to bacteriology. *Journal of Bacteriology*, 23: 1–10

Brucker RM, Bordenstein SR (2012) Speciation by symbiosis. *Trends in Ecology and Evolution* 27: 443–451

Brückner S, H-U Mösch (2012) Choosing the right lifestyle: adhesion and development in *Saccharomyces cerevisiae*. *FEMS Microbiology Reviews* 36: 25–58

Bryant DA, Frigaard N-U (2006) Prokaryotic photosynthesis and phototrophy illuminated. *Trends in Microbiology* 14: 488–496

Bryant RS (1987) Potential uses of microorganisms in petroleum recovery technology. *Proceedings of the Oklahoma Academy of Science* 67: 97–104

Buckling A, Maclean RC, Brockhurst MA, Colegrave N (2009) The *Beagle* in a bottle. *Nature* 457: 824–829

Buick R (2008) When did oxygenic photosynthesis evolve? *Philosophical Transactions of the Royal Society of London B* 363: 2731–2743

Burke C, Steinberg P, Rusch D, Kjelleberg S, Thomas T (2011) Bacterial community assembly based on functional genes rather than species. *Proceedings of the National Academy of Sciences USA* 108: 14288–14293

Cabeen MT, Jacobs-Wagner C (2010) The bacterial cytoskeleton. *Annual Review of Genetics* 44: 365–392

Cadillo-Quiroz H, Didelot X, Held NL, Herrera A, Darling A, Reno ML, Krause DJ, Whitaker RJ (2012) Patterns of gene flow define species of thermophilic archaea. *PLoS Biology* 10(2):e1001265, doi:10.1037/journal.pbio.1001265

Cairns J, Overbaugh J, Miller S (1988). The origin of mutants. *Nature* 335: 142–145

Calcott B, Sterelny K (2011) Introduction: a dynamic view of evolution. In Calcott B, Sterelny K (eds), *The Major Transitions in Evolution Revisited* (pp. 1–14) Cambridge, MA: MIT Press

Canfield DE (2005) The early history of atmospheric oxygen: homage to Robert M. Garrels. *Annual Review of Earth Planetary Science* 33: 1–36

Canfield DE, Rosing MT, Bjerrum C (2006) Early anaerobic metabolisms. *Philosophical Transactions of the Royal Society London B* 361: 1819–1836

Caporaso JG, Lauber CL, Costello EK, Berg-Lyons D, Gonzalez A, Stombaugh J, Knights D, Gajer P, Ravel J, Fierer N, et al. (2011) Moving pictures of the human microbiome. *Genome Biology* 12:R50, doi:10.1186/gb-2011-12-5-r50

Caro-Quintero A, Konstantinidis KT (2012) Bacterial species may exist, metagenomics reveal. *Environmental Microbiology* 14: 347–355

Caron DA, Davis PG, Sieburth JMcN (1989) Factors responsible for the differences in cultural estimates and direct microscopical counts of populations of bacterivorous nanoflagellates. *Microbial Ecology* 18: 89–104

Carroll SB (2008) Evo-devo and an expanding evolutionary synthesis: a genetic theory of morphological evolution. *Cell* 134: 25–36

Caruso T, Chan Y, Lacap DC, Lau MCY, McKay CP, Pointing SB (2011) Stochastic and deterministic processes interact in the assembly of desert microbial communities on a global scale. *ISME Journal* 5: 1406–1413

Cavalier-Smith T (1987) The origin of eukaryote and archaebacterial cells. *Annals of the New York Academy of Sciences* 503: 17–54

—(2002) The phagotrophic origin of eukaryotes and phylogenetic classification of Protozoa. *International Journal of Systematic and Evolutionary Microbiology* 52: 297–354

—(2004) Only six kingdoms of life. *Proceedings of the Royal Society London B* 271: 1251–1262

—(2007) Concept of a bacterium still valid in prokaryote debate. *Nature* 446: 257

—(2010) Deep phylogeny, ancestral groups and the four ages of life. *Philosophical Transactions of the Royal Society London B* 365: 111–132

—(2013) Early evolution of eukaryote feeding modes, cell structural diversity, and classification of the protozoan phyla Loukosoa, Sulcozoa, and Choanozoa. *European Journal of Protistology* 49: 115–178

Cavalli-Sforza LL (1992) Forty years ago in genetics: the unorthodox mating behavior of bacteria. *Genetics* 132: 635–637

Cavendish M (2001 [1666, 1668]) *Observations Upon Mechanical Philosophy*. In O'Neill E (ed), *Observations Upon Mechanical Philosophy*. Cambridge University Press.

Celiker H, Gore J (2013) Cellular cooperation: insights from microbes. *Trends in Cell Biology* 23: 9–15

Chao L, Levin BR, Stewart FM (1977) A complex community in a simple habitat: an experimental study with bacteria and phage. *Ecology* 58: 369–378

Chen WW, Niepel M, Sorger PK (2010) Classic and contemporary approaches to modeling biochemical reactions. *Genes and Development* 24: 1861–1875

Cho I, Yaminishi S, Cox L, Methé BA, Zavadil J, Li K, Gao Z, Mahana D, Raju K, Teitler I, Li H, Alekseyenko AV, Blaser MJ (2012) Antibiotics in early life alter the murine colonic microbiome and adiposity. *Nature* 488: 621–626

Ciccarelli FD, Doerks T, von Mering C, Creevey CJ, Snel B, Bork P (2006) Toward automatic reconstruction of a highly resolved tree of life. *Science* 311: 1283–1287

Claverie J-M (2006) Viruses take center stage in cellular evolution. *Genome Biology* 7: 110, doi:10.1186/gb-2006-7-6-110

Claverie J-M, Ogata H (2009) Ten good reasons not to exclude viruses from the evolutionary picture. *Nature Reviews Microbiology* 7: 615, doi:10.1038/nrmicro2108-c3

Cleland CE (2013) Pluralism or unity in biology: could microbes hold the secret to life? *Biology and Philosophy* 28: 189–204

Cleland CE, Chyba CF (2002) Defining 'life'. *Origins of Life and Evolution of the Biosphere* 32: 387–393

Clements FE (1916) *Plant Succession: An Analysis of the Development of Vegetation.* Washington, DC: Carnegie Institution. Available online at: http://archive.org/details/cu31924000531818

Cockell CS (2005) The value of microorganisms. *Environmental Ethics* 27: 375–390

—(2008) Environmental ethics and size. *Ethics and the Environment* 13: 23–39

Cockell CS, Jones HL (2009) Advancing the case for microbial conservation. *Oryx* 43: 520–526

Cohan FM (2002) What are bacterial species? *Annual Review of Microbiology* 56: 457–487

—(2006) Towards a conceptual and operational union of bacterial systematics, ecology, and evolution. *Philosophical Transactions of the Royal Society London B* 361: 1985–1996

Cohen O, Gophna U, Pupko T (2011) The complexity hypothesis revisited: connectivity rather than function constitutes a barrier to horizontal gene transfer. *Molecular Biology and Evolution* 28: 1481–1489

Cole LJ, Wright WH (1916) Application of the pure-line concept to bacteria. *Journal of Infectious Diseases* 19: 209–221

Colegrave N, Collins S (2008) Experimental evolution: experimental evolution and evolvability. *Heredity* 100: 464–470

Colwell RR (1997) Microbial diversity: the importance of exploration and conservation. *Journal of Industrial Microbiology & Biotechnology* 18: 302–307

Committee on Metagenomics (2007) *The New Science of Metagenomics: Revealing the Secrets of our Microbial Planet.* Washington, DC: National Academies Press

Conway Morris S (1998) The evolution of diversity in ancient ecosystems: a review. *Philosophical Transactions of the Royal Society B* 353: 327–345

Copeland HF (1938) The kingdoms of organisms. *Quarterly Review of Biology* 13: 383–420

Corliss JO (1984) The kingdom Protista and its 45 phyla. *BioSystems* 17: 87–126

Costello EK, Stagaman K, Dethlefsen L, Bohannan BJM, Relman DA (2012) The application of ecological theory toward and understanding of the human microbiome. *Science* 336: 1255–1262

Cowan ST (1971) Sense and nonsense in bacterial taxonomy. *Journal of General Microbiology* 67: 1–8

Cox CJ, Foster PG, Hirt RP, Harris SR, Embley TM (2008) The archaebacterial origin of eukaryotes. *Proceedings of the National Academy of Sciences USA* 105: 20356–20361

Crawford JW, Harris JA, Ritz K, Young IM (2005) Towards an evolutionary ecology of life in soil. *Trends in Ecology and Evolution* 20: 81–87

Creager ANH (2001) *The Life of a Virus: Tobacco Mosaic Virus as an Experimental Model, 1930–1965.* Chicago: Chicago University Press

Cryan JF, Dinan TG (2012) Mind-altering microorganisms: the impact of the gut microbiota on brain and behaviour. *Nature Reviews Neuroscience* 13: 701–712

Curtis TP (2006) Microbial ecologists: it's time to 'go large'. *Nature Reviews Microbiology* 4: 488

—(2007) Theory and the microbial world. *Environmental Microbiology* 9: 1 (Crystal Ball section)

Curtis TP, Sloan W (2005) Exploring microbial diversity – a vast below. *Science* 309: 1331–1333

Curtis TP, Head IM, Lunn M, Woodcock S, Schloss PD, Sloan WT (2006) What is the extent of prokaryotic diversity? *Philosophical Transactions of the Royal Society London B* 361: 2023–2037

D'Costa V, King CE, Kalan L, Morar M, Sung WWL, Schwartz C, Froese D, Zazula G, Calmels F, Debruyne R, et al. (2011) Antibiotic resistance is ancient. *Nature* 477: 457–461

Dagan T, Martin W (2006) The tree of one percent. *Genome Biology* 7:118, doi:10.1186/gb-2006-7-10-118

—(2010) Getting a better picture of microbial evolution en route to a network of genomes. *Philosophical Transactions of the Royal Society B* 364: 2187–2196

Dallinger WH (1878) Letter to C. R. Darwin (29 June, 1878, Letter 11576), www.darwinproject.ac.uk

Danovaro R, Dell-Anno A, Corinaldesi C, Maganini M, Noble R, Tambourini C, Weinbauer M (2008) Major viral impact on the functioning of benthic deep-sea ecosystems. *Nature* 454: 1084–1086

Dantas G, Sommer MOA, Degnan PH, Goodman AL (2013) Experimental approaches for defining functional roles of microbes in the human gut. *Annual Review of Microbiology* 67: 459–475

Dantas G, Sommer MOA, Oluwasegun RD, Church GM (2008) Bacteria subsisting on antibiotics. *Science* 320: 100–103

Darch SE, West SA, Winzer K, Diggle SP (2012) Density-dependence fitness benefits in quorum-sensing bacterial populations. *Proceedings of the National Academy of Sciences USA* 109: 8259–8263

Darden L, Tabery J (2009) Molecular biology. In Zalta EN (ed), *Stanford Encyclopedia of Philosophy*, http://plato.stanford.edu/entries/molecular-biology/

Darwin CR (1861) *On the Origin of Species by Means of Natural Selection, or the Preservation of Favoured Races in the Struggle for Life* (3rd ed). London: John Murray

Darwin F (ed) (1887) *The Life and Letters of Charles Darwin*, Vol 2. London: John Murray

Davidson EH (2010) Emerging properties of animal gene regulatory networks. *Nature* 468: 911–920

Davies J (2010) Anthropomorphism in science. *EMBO reports* 11: 721

Davies J, Davies D (2010) Origins and evolution of antibiotic resistance. *Microbiology and Molecular Biology Reviews* 74: 417–433

Davis RH (2003) *The Microbial Models of Molecular Biology: From Genes to Genomes*. Oxford: Oxford University Press

Dawkins R (2006 [1976]) *The Selfish Gene* (30th anniversary ed). Oxford: OUP

Day MD, Beck D, Foster JA (2011) Microbial communities as experimental units. *BioScience* 61: 398–406

de Duve C (2007) The origin of eukaryotes: a reappraisal. *Nature Reviews Genetics* 8: 395–403

de Hoon MJL, Eichenberger P, Vitkup D (2010) Hierarchical evolution of the bacterial sporulation network. *Current Biology* 20: R735–R745

de Lorenzo V, Pieper D, Ramos JL (2013) From the test tube to the environment – and back. *Environmental Microbiology* 15: 6–11

de Queiroz K (1997) The Linnaean hierarchy and the evolutionization of taxonomy, with emphasis on the problem of nomenclature. *Aliso* 15: 125–144

de Visser JAGM, Elena SF (2007) The evolution of sex: empirical insights into the roles of epistasis and drift. *Nature Reviews Genetics* 8: 139–149

de Wit R, Bouvier T (2006) 'Everything is everywhere, but, the environment selects'; what did Baas Becking and Beijerinck really say? *Environmental Microbiology* 8: 755–758

Delbrück M (2000 [1949]) A physicist looks at biology: Address delivered at the 100th meeting of the Academy. In Cairns J, Stent GS, Watson JD (eds), *Phage and the Origins of Molecular Biology* (pp. 9–22). Cold Spring Harbor: Cold Spring Harbor Press

—(1969) Nobel lecture, Dec 10, 1969. Available at: http://nobelprize.org/nobel_prizes/medicine/laureates/1969/delbruck-lecture.html

Delbrück M, Bailey WT Jr (1946) Induced mutations in bacterial viruses. *Cold Spring Harbor Symposia on Quantitative Biology* 11: 33–37

Delwiche CF, Timme RE (2011) Primer: plants. *Current Biology* 21: R417–R422

Demerec M (1946) Minutes of the 23rd Annual Meeting of the Long Island Biological Association, July 30th, 1946. Cold Spring Harbor Archives (with special thanks to archivist Clare Clark)

Denamur E, Matic I (2006) Evolution of mutation rates in bacteria. *Molecular Microbiology* 60: 820–827

Denison RF, Kiers ET (2011) Life histories of symbiotic rhizobia and mycorrhizal fungi. *Current Biology* 21: R775–R785

Dethlefsen L, Relman DA (2011) Incomplete recovery and individualized responses of the human distal gut microbiota to repeated antibiotic perturbation. *Proceedings of the National Academy of Sciences USA* 108: 4554–4561

Dethlefsen L, McFall-Ngai M, Relman DA (2007) An ecological and evolutionary perspective on human-microbe mutualism and disease. *Nature* 449: 811–818

Diamond JM (1983) Laboratory, field and natural experiments. *Nature* 304: 586–587

Dietrich LEP, Tice MM, Newman DK (2006) The co-evolution of life and Earth. *Current Biology* 16: R395–R400

Diggle SP, Gardner A, West SA, Griffin AS (2007) Evolutionary theory of bacterial quorum sensing: when is a signal not a signal? *Philosophical Transactions of the Royal Society London B* 362: 1241–1249

Dini-Andreote F, van Elsas JD (2013) Back to the basics: the need for ecophysiological insights to enhance our understanding of microbial behaviour in the rhizosphere. *Plant Soil* 373: 1–15

Dixon B (1976) Smallpox, imminent extinction and an unresolved dilemma. *New Scientist* 26: 430–432

Dobell C (1932) *Antony van Leeuwenhoek and his 'Little Animals': Being Some Account of the Father of Protozoology and Bacteriology and his Multifarious Discoveries in these Disciplines.* NY: Harcourt, Brace & Co. Available online at: www.archive.org/details/antonyvanleeuwenoodobe

Dobzhansky TG (1941) *Genetics and the Origin of Species* (2nd ed). NY: Colombia

—(1973) Nothing in biology makes sense except in the light of evolution. *American Biology Teacher* 35: 125–129

Doney SC (1999) Major challenges confronting marine biogeochemical modeling. *Global Biogeochemical Cycles* 13: 705–714

Doney SC, Abbott RM, Cullen JJ, Karl DM, Rothstein L (2004) From genes to ecosystems: the ocean's new frontier. *Frontiers in Ecology and the Environment* 2: 457–466

Doolittle WF (1999) Phylogenetic classification and the universal tree. *Science* 284: 2124–2129

Doolittle WF, Papke RT (2006) Genomics and the bacterial species problem. *Genome Biology* 7: 116, doi:10.1186/gb-2006-7-9-116

Doolittle WF, Zhaxybayeva O (2009) On the origin of prokaryotic species. *Genome Research* 19: 744–756

—(2010) Metagenomics and the units of biological organization. *BioScience* 60: 102–112

—(2013) What is a prokaryote? In Rosenberg E, DeLong EF, Lory S, Stackebrandt E, Thompson T (eds), *The Prokaryotes: Prokaryotic Biology and Symbiotic Associations* (pp. 21–37). Berlin: Springer-Verlag

Douglas AE (2008) Conflicts, cheats and the persistence of symbioses. *New Phytologist* 177: 849–858

Dretske F (1986) Misrepresentation. In Bogdan RJ (ed), *Belief: Form, Content, and Function* (pp. 17–36). Oxford: Clarendon

Drews G (2000) The roots of microbiology and the influence of Ferdinand Cohn on microbiology of the 19th century. *FEMS Microbiology Reviews* 24: 225–249

Dubilier N, Bergin C, Lott C (2008) Symbiotic diversity in marine animals: the art of harnessing chemosynthesis. *Nature Reviews Microbiology* 6: 725–740

Ducklow H (2008) Microbial services: challenges for microbial ecologists in a changing world. *Aquatic Microbial Ecology* 53: 13–19

Duclaux E (1920) *Louis Pasteur: The History of a Mind* (transl. Smith EF, Hedges F). Philadelphia: W. B. Saunders. Available online at: http://pasteurbrewing.com/biography/biography/the-history-of-a-mind-by-emile-duclaux.html

Dupré J (1993) *The Disorder of Things: Metaphysical Foundations of the Disunity of Science*. Cambridge, MA: Harvard University Press

Dupré J , O'Malley MA (2007) Metagenomics and biological ontology. *Studies in History and Philosophy of Biological and Biomedical Sciences* 38: 834–846

—(2009) Varieties of living things: life at the intersection of lineage and metabolism. *Philosophy and Theory in Biology* 1: e003, http://hdl.handle.net/2027/spo.6959004.0001.003

Dupressoir A, Lavialle C, Heidmann T (2012) From ancestral infectious retroviruses to bona fide cellular genes: role of the captured *syncytins* in placentation. *Placenta* 33: 663–671

Dworkin M (2012) Sergei Winogradsky: a founder of modern microbiology and the first microbial ecologist. *FEMS Microbiology Reviews* 36: 364–379

Dykhuizen DE, Davies M (1980) An experimental model: bacterial specialists and generalists competing in chemostats. *Ecology* 61: 1213–1227

Dykhuizen DE (1990) Experimental studies of natural selection in bacteria. *Annual Review of Ecology and Systematics* 21: 373–398

—(1993) Chemostats used for studying natural selection and adaptive evolution. *Methods in Enzymology* 224: 613–631

Dykhuizen DE, Green L (1991) Recombination in *Escherichia coli* and the definition of biological species. *Journal of Bacteriology* 173: 7257–7268

Dykhuizen DE, Hartl DL (1983) Selection in chemostats. *Microbiology Reviews* 47: 150–168

Earle DJ, Deem MW (2004) Evolvability is a selectable trait. *Proceedings of the National Academy of Sciences USA* 101: 11531–11536

Ebersberger I, de Matos Simoes R, Kupczok A, Gube M, Kothe E, Voigt K, von Haeseler A (2012) A consistent phylogenetic backbone for the fungi. *Molecular Biology and Evolution* 29: 1319–1334

Editorial (1869) The origin of life. *British Medical Journal* 1 (431): 312–313

—(2011) Microbiology by the numbers. *Nature Reviews Microbiology* 9: 628

Ehrenberg CG (1838) *Die Infusionsthierchen als Volkommene Organismen: Ein Blick in das Tiefere Organische Leben der Natur*. Leipzig: Voss

Eigen M (1993) Viral quasispecies. *Scientific American* July: 42–49

Eldar A, Chary VK, Xenopoulos P, Fontes ME, Losón, Dworkin J, Piggot PJ, Elowitz MB (2009) Partial penetrance facilitates developmental evolution in bacteria. *Nature* 460: 510–514

Elena SF, Lenski RE (2003) Evolution experiments with microorganisms: the dynamics and genetic bases of adaptation. *Nature Reviews Genetics* 4: 457–469

Embley TM (2012) SMBE Plenary Lecture: Unraveling the chimeric origins of eukaryotes: trees, genomes and organelles. SMBE2012, Dublin (June 23–26), www.smbe2012.org/scientific-content/keynote-speakers.html

Embley TM, Martin W (2006) Eukaryotic evolution, changes and challenges. *Nature* 440: 623–630

Ereshefsky M (1991) Species, higher taxa, and the units of evolution. *Philosophy of Science* 58: 84–101

—(2008) Systematics and taxonomy. In Sarkar S, Plutynski A (eds), *A Companion to Philosophy of Biology* (pp. 99–118). Malden, MA: Blackwell

—(2010) Microbiology and the species problem. *Biology and Philosophy* 25: 553–568

Ezenwa VO, Gerardo NM, Inouye DW, Medina M, Xavier JB (2012) Animal behavior and the microbiome. *Science* 338: 198–199

Faith DP (2007) Biodiversity. In Zalta EN (ed), *Stanford Encyclopedia of Philosophy*, http://plato.stanford.edu/entries/biodiversity

Fajardo A, Martínez JL (2008) Antibiotics as signals that trigger specific bacterial responses. *Current Opinion in Microbiology* 11: 161–167

Falkowski PG (2006) Tracing oxygen's imprint on Earth's metabolic evolution. *Science* 311: 1724–1725

Falkowski PG, Godfrey LV (2008) Electrons, life and the evolution of Earth's oxygen cycle. *Philosophical Transactions of the Royal Society London B* 363: 2705–2716

Falkowski PG, Isozaki Y (2008) The story of O_2. *Science* 322: 540–542

Falkowski PG, Fenchel T, DeLong E (2008) The microbial engines that drive Earth's biogeochemical cycles. *Science* 320: 1034–1039

Farley J (1974) *The Spontaneous Generation Controversy from Descartes to Oparin.* Baltimore: Johns Hopkins

Ferenci T (2008) Bacterial physiology, regulation and mutational adaptation in a chemostat environment. *Advances in Microbial Physiology* 53: 169–229

Feschotte C, Gilbert C (2012) Endogenous viruses: insights into viral evolution and impact on host biology. *Nature Reviews Genetics* 13: 283–296

Field KG, Olsen GJ, Lane DJ, Giovannoni SJ, Ghiselin MT, Raff EC, Pace NR, Raff RA. (1988) Molecular phylogeny of the animal kingdom. *Science* 239: 748–753

Fierer N, Jackson RB (2006) The diversity and biogeography of soil bacterial communities. *Proceedings of the National Academy of Sciences USA* 103: 626–631

Fierer N, Ferrenberg S, Flores GE, González A, Kueneman J, Legg T, Lynch RC, McDonald D, Mihaljevic JR, O'Neill SP, et al. (2012a) From animalcules to an ecosystem: application of ecological concepts to the human microbiome. *Annual Review of Ecology, Evolution, and Systematics* 43: 137–155

Fierer N, Leff JW, Adams BJ, Nielsen UN, Bates ST, Lauber CL, Owens S, Gilbert JA, Wall DH, Caporaso JG (2012b) Cross-biome metagenomic analyses of soil microbial communities and their functional attributes. *Proceedings of the National Academy of Sciences USA* 109: 21390–21395

Finkel SE (2006) Long-term survival during stationary phase: evolution and the GASP phenotype. *Nature Reviews Microbiology* 4: 113–120

Finlay BJ, Esteban GF, Brown S, Fenchel T, Hoef-Emden K (2006) Multiple cosmopolitan ecotypes within a microbial eukaryote morphospecies. *Protist* 157: 377–390

Flores E, Herrero A (2010) Compartmentalized function through cell differentiation in filamentous cyanobacteria. *Nature Reviews Microbiology* 8: 39–50

Fodor JA (1986) Why paramecia don't have mental representations. *Midwest Studies in Philosophy* 10: 3–23

Forterre P (2006) Three RNA cells for ribosomal lineages and three DNA viruses to replicate their genomes: a hypothesis for the origin of cellular domain. *Proceedings of the National Academy of Sciences USA* 103: 3669–3674

—(2013) The virocell concept and environmental microbiology. *ISME Journal* 7: 233–236

Foster KR (2011) The sociobiology of molecular systems. *Nature Reviews Genetics* 12: 193–203

Foster KR, Parkinson K, Thompson CRL (2006) What can microbial genetics teach sociobiology? *Trends in Genetics* 23: 74–80

Francis CA, Beman JM, Kuypers MMM (2007) New processes and players in the nitrogen cycle: the microbial ecology of anaerobic and archaeal ammonia oxidation. *ISME Journal* 1: 19–27

Frank SA (2003) Repression of competition and the evolution of cooperation. *Evolution* 57: 693–705

Frankel RB (2009) The discovery of magnetotactic/magnetosensitive bacteria. *Chinese Journal of Oceanology and Limnology* 27: 1–2

Frankel RB, Blakemore RP, de Araujo FFT, Danon DMSEJ (1981) Magnetotactic bacteria at the geomagnetic equator. *Science* 212: 1269–1270

Fraser C, Hanage WP, Spratt BG (2007) Recombination and the nature of bacterial speciation. *Science* 315: 476–480

Fraser C, Hanage WP, Spratt BG, Alm EJ, Polz MF, Spratt BG, Hanage WP (2009) The bacterial species challenge: making sense of genetic and ecological diversity. *Science* 323: 741–746

Fraune S, Bosch TCG (2010) Why bacteria matter in animal development and evolution. *BioEssays* 32: 571–580

Friedmann HC (2004) From '*Butyribacterium*' to '*E. coli*': an essay on unity on biochemistry. *Perspectives in Biology and Medicine* 47: 47–66

Fry I (2010) The role of natural selection in the origin of life. *Origins of Life and Evolution of Biospheres* 41: 3–16

Fuerst JA (2005) Intracellular compartmentation in Planctomycetes. *Annual Review of Microbiology* 59: 299–328

Fuhrman JA (1999) Marine viruses and their biogeochemical and ecological effects. *Nature* 399: 541–548

Furuya EY, Lowy FD (2006) Antimicrobial-resistant bacteria in the community setting. *Nature Reviews Microbiology* 4, 36–45

Gardner A, Grafen A (2009) Capturing the superorganism: a formal theory of group adaptation. *Journal of Evolutionary Biology* 22: 659–671

Gause GF (1932) Experimental studies on the struggle for existence. *Journal of Experimental Biology* 9: 389–402

—(1934) *The Struggle for Existence*. Baltimore MD: Williams and Wilkins

Gest H (2004) The discovery of microorganisms by Robert Hooke and Antoni Van Leeuwenhoek, Fellows of the Royal Society. *Notes and Records of the Royal Society London* 58: 187–201

Gevers D, Cohan FM, Lawrence JG, Spratt BG, Coenye T, Feil EJ, Stackebrandt E, de Peer YV, Vandamme P, Thompson FL, Swings J (2005) Re-evaluating prokaryotic species. *Nature Reviews Microbiology* 3: 733–739

Ghiglione J-F, Galand PE, Pommier T, Pedrós-Alió C, Maas EW, Bakker K, Bertilson S, Kirchman DL, Lovejoy C, Yager PL, Murray AE (2012) Pole-to-pole biogeography of surface and deep marine bacterial communities. *Proceedings of the National Academy of Sciences USA* 109: 17633–17638

Gibson AH (2013) Edward O. Wilson and the organicist tradition. *Journal of the History of Biology* 46: 599–630

Gilbert JA, Dupont CL (2011) Microbial metagenomics: beyond the genome. *Annual Review of Marine Science* 3: 347–371

Gilbert JA, Steele JA, Caporaso JG, Steinbrück L, Reeder J, Temperton B, Huse S, McHardy AC, Knight R, Joint I, Somerfield P, Fuhrman JA, Field D (2012) Defining seasonal marine microbial community dynamics. *ISME Journal* 6: 298–308

Godfrey-Smith P (2001) Three kinds of adaptationism. In Orzack SH, Sober E (eds), *Adaptationism and Optimality* (pp. 335–357). Cambridge: Cambridge University Press

—(2002) Complexity and cognition. In Sternberg R, Kaufman J (eds), *The Evolution of Intelligence* (pp. 233–249). Mahwah: Lawrence Erlbaum

—(2006) Mental representation, naturalism, and teleosemantics. In Macdonald G, Papineau D (eds), *Teleosemantics: New Philosophical Essays* (pp. 42–68). Oxford: Clarendon

—(2009) *Darwinian Populations and Natural Selection.* Oxford: Oxford University Press

—(2013) Darwinian individuals. In Bouchard F, Huneman P (eds), *From Groups to Individuals: Perspectives on Biological Associations and Emerging Individuals* (pp. 17–36). Cambridge, MA: MIT Press

Gleason HA (1926) The individualistic concept of the plant association. *Bulletin of the Torrey Botanical Club* 53: 7–26

Gogarten JP, Doolittle WF, Lawrence JG (2002) Prokaryotic evolution in light of gene transfer. *Molecular Biology and Evolution* 19: 2226–2238

Gogarten JP, Doolittle WF, Lawrence JG, Townsend JP (2005) Horizontal gene transfer, genome innovation and evolution. *Nature Reviews Microbiology* 3: 679–687

Goldenfeld N, Woese C (2011) Life is physics: evolution as a collective phenomenon far from equilibrium. *Annual Review of Condensed Matter Physics* 2: 375–399.

Gomez-Alvarez V, Teal TK, Schmidt TM (2009) Systematic artifacts in metagenomes from complex microbial communities. *ISME Journal* 3: 1314–1317

Goo E, Majerczyk CD, An JH, Chandler JR, Seo Y-S, Ham H, Lim JY, Kim H, Lee B, Jang MS, et al. (2012) Bacterial quorum sensing, cooperativity, and anticipation of stationary-phase stress. *Proceedings of the National Academy of Sciences USA* 109: 19775–19780

Gorelick R, Heng HHQ (2010) Sex reduces genetic variation: a multidisciplinary review. *Evolution* 65: 1088–1098

Gould SJ (1978) The exaptive excellence of spandrels as a term and prototype. *Proceedings of the National Academy of Sciences USA* 94: 10750–10755

—(1979) A natural precision designer. *New Scientist* November 8th: 447–448

—(1980) Natural attraction: bacteria, the birds and the bees. In SJ Gould, *The Panda's Thumb: More Reflections on Natural History* (pp. 306–314). NY: WW Norton

—(1994) The evolution of life on earth. *Scientific American* 271: 84–91

—(1997) *Life's Grandeur: The Spread of Excellence from Plato to Darwin.* London: Vintage

Gould SJ, Lewontin RC (1979) The spandrels of San Marco and the Panglossian paradigm: a critique of the adaptationist programme. *Proceedings of the Royal Society London B* 205: 581–598

Gradmann C (2000) Isolation, contamination, and pure culture: monomorphism and polymorphism of pathogenic microorganisms as research problem 1860–1880. *Perspectives on Science* 9: 147–172

Grande FD, Widmer I, Wagner HH, Scheidegger C (2012) Vertical and horizontal photobiont transmission within populations of a lichen symbiosis. *Molecular Ecology* 21: 3159–3172

Green JL, Holmes AJ, Westoby M, Oliver I, Briscoe D, Dangerfield M, Gillings M, Beattie AJ (2004) Spatial scaling of microbial eukaryote diversity. *Nature* 432: 747–750

Green JL, Bohannan BJM, Whitaker RJ (2008) Microbial biogeography: from taxonomy to traits. *Science* 320: 1039–1042

Grene M, Depew D (2004) *The Philosophy of Biology: An Episodic History.* Cambridge: Cambridge University Press

Gribaldo S, Brochier-Armanet C (2012) Time for order in microbial systematics. *Trends in Microbiology* 20: 209–210

Griesemer J (2000) The units of evolutionary transition. *Selection* 1: 67–80

Grosberg RK, Strathmann RR (1998) One cell, two cell, red cell, blue cell: the persistence of a unicellular stage in multicellular life histories. *Trends in Ecology and Evolution* 13: 112–116

—(2007) The evolution of multicellularity: a minor major transition? *Annual Review of Ecology, Evolution, and Systematics* 38: 621–654

Grundmann GL (2004) Spatial scales of soil bacterial diversity – the size of a clone. *FEMS Microbiology Ecology* 48: 119–127

Guljamow A, Jenke-Kodama H, Saumweber H, Quillardet P, Frangeul L, Castets AM, Bouchier C, de Marsac NT, Dittmann E (2007) Horizontal gene transfer of two cytoskeletal elements from a eukaryote to a cyanobacterium. *Current Biology* 17: R757–759

Guo FF, Yang W, Jiang W, Geng S, Peng T, Li JL (2012) Magnetosomes eliminate reactive oxygen species in *Magnetospirillum gryphiswaldense* MSR-1. *Environmental Microbiology* 14: 1722–1729

Gupta RS (1998) What are archaebacteria: life's third domain or monoderm prokaryotes related to Gram-positive bacteria? A new proposal for the classification of prokaryotic organisms. *Molecular Microbiology* 29: 695–707

Haber M (2013) Colonies are individuals: revisiting the superorganism revival. In Bouchard F, Huneman P (eds), *From Groups to Individuals: Evolution and Emerging Individuality* (pp. 195–217). Cambridge, MA: MIT Press

Hacking I (1983) *Representing and Intervening: Introductory Topics in the Philosophy of Natural Science*. Cambridge: Cambridge University Press

Hadley P (1937) Further advances in the study of microbic dissociation. *Journal of Infectious Diseases* 60: 129–192

Haeckel E (1869a) Monograph of Monera (Part 3). *Quarterly Journal of Microscopical Science* s2–9: 219–232

—(1869b) Monograph of Monera (Part 4). *Quarterly Journal of Microscopical Science* s2–9: 327–342

Hagen JB (2012) Five kingdoms, more or less: Robert Whittaker and the broad classification of organisms. *BioScience* 62: 67–74

Hajishengallis G, Darveau RP, Curtis MA (2012) The keystone-pathogen hypothesis. *Nature Reviews Microbiology* 10: 717–725

Hamady M, Knight R (2009) Microbial community profiling for human microbiome projects: tools, techniques, and challenges. *Genome Research* 19: 1141–1152

Hanage WP, Fraser C, Spratt BG (2005) Fuzzy species among recombinogenic bacteria. *BMC Biology* 3:6, doi:10.1186/1741-7007-3-6

Handelsman J, Rondon MR, Brady SF, Clardy J, Goodman RM (1998) Molecular biological access to the chemistry of unknown soil microbes: a new frontier for natural products. *Chemical Biology* 5: R245–49

Hanson CA, Fuhrman JA, Horner-Devine MC, Martiny JBH (2012) Beyond biogeographic patterns: processes shaping the microbial landscape. *Nature Reviews Microbiology* 10: 497–506

Hardin G (1960) The competitive exclusion principle. *Science* 131: 1292–1297

Harrison E, Brockhurst MA (2012) Plasmid-mediated horizontal gene transfer is a coevolutionary process. *Trends in Microbiology* 20: 262–267

Harrison JP, Gheeraert N, Tsigelnitskiy D, Cockell CS (2013) The limits for life under multiple extremes. *Trends in Microbiology* 21: 204–212

Hayes W (1953) Observations on a transmissible agent determining sexual differentiation in *Bacterium coli*. *Journal of General Microbiology* 8: 72–88

—(1966) The Leeuwenhoek Lecture, 1965: Some controversial aspects of bacteria sexuality. *Proceedings of the Royal Society London B* 165: 1–19

Hazelbauer GL (2012) Bacterial chemotaxis: the early years of molecular studies. *Annual Review of Microbiology* 66: 285–303

Heijtz RD, Wang S, Anuar F, Qian Y, Björkholm B, Samuelsson A, Hibberd ML, Forssberg H, Pettersson S (2011) Normal gut microbiota modulates brain development and behavior. *Proceedings of the National Academy of Sciences USA* 108: 3047–3052

Hekstra DR, Leibler S (2012) Contingency and statistical laws in replicate microbial closed ecosystems. *Cell* 149: 1164–1173

Henry LM, Peccoud J, Simon J-C, Hadfield JD, Maiden MJC, Ferrari J, Godfray HCJ (2013) Horizontally transmitted symbionts and host colonization of ecological niches. *Current Biology* 23: 1713–1717

Herre EA, Knowlton N, Mueller UG, Rehner SA (1999) The evolution of mutualisms: exploring the paths between cooperation and conflict. *Trends in Ecology and Evolution* 14: 49–53

Hershey AD (1946) Spontaneous mutations in microorganisms. *Cold Spring Harbor Symposia on Quantitative Biology* 11: 67–77

Hindré T, Knibbe C, Beslon G, Schneider D (2012) New insights into bacterial adaptation through *in vivo* and *in silico* experimental evolution. *Nature Reviews Microbiology* 10: 352–365

Hodkinson BP, Gottel NR, Schadt CW, Lutzoni F (2011) Photoautotrophic symbiont and geography are major factors affecting highly structured and diverse bacterial communities in the lichen microbiome. *Environmental Microbiology* 14: 147–161

Hoehler TM, Jørgensen BB (2013) Microbial life under extreme energy limitation. *Nature Reviews Microbiology* 11: 83–94

Hohmann-Marriott MF, Blankenship RE (2011) Evolution of photosynthesis. *Annual Review of Plant Biology* 62: 515–548

Holmes EC (2011) What does virus evolution tell us about virus origins? *Journal of Virology* 85: 5247–5251

Hooke R (1665) *Micrographia: or Some Physiological Descriptions of Minute Bodies Made by Magnifying Glasses with Observations and Inquiries Thereupon.* London: Martyn & Allestry, Printers for the Royal Society of London. Available online at: http://www.gutenberg.org/ebooks/15491

Hooper LV, Littman DR, Macpherson AJ (2012) Interactions between the microbiota and the immune system. *Science* 336: 1268–1273

Horner-Devine MC, Lage M, Hughes JB, Bohannan BJM (2004) A taxa-area relationship for bacteria. *Nature* 432 750–753

Hottopp JCD (2011) Horizontal gene transfer between bacteria and animals. *Trends in Genetics* 27: 157–163

Howe CJ, Barbrook AC, Nisbet RER, Lockhart PJ, Larkum AWD (2008) The origin of plastids. *Philosophical Transactions of the Royal Society London B* 363: 2675–2685

Hubbell SP (2001) *The Unified Neutral Theory of Biodiversity and Biogeography.* Princeton, NJ: Princeton University Press

Huber JA, Welch DBM, Morrison HG, Huse SM, Neal PR, Butterfield DA, Sogin ML (2007) Microbial population structures in the deep marine biosphere. *Science* 318: 97–100

Hungate RE (1960) Microbial ecology of the rumen. *Journal of Bacteriology* 24: 353–364

—(1979) Evolution of a microbial ecologist. *Annual Review of Microbiology* 33: 1–20

Huxley J (1942) *Evolution: The Modern Synthesis.* London: George Allen & Unwin

—(2010 [1974]) *Evolution: The Modern Synthesis. The Definitive Edition.* Cambridge, MA: MIT Press

Ingham CJ, Kalisman O, Finkelshtein A, Ben-Jacob E (2011) Mutually facilitated dispersal between the nonmotile fungus *Aspergillus fumigatus* and the swarming bacterium *Paenibacillus vortex*. *Proceedings of the National Academy of Sciences USA* 108: 19731–19736

Ivanov S, Federova EE, Limpens E, De Mita S, Genre A, Bonfante P, Bisseling T (2012) Rhizobium-legume symbiosis shares an exocytotic pathway required for arbuscule formation. *Proceedings of the National Academy of Sciences USA* 109: 8316–8321

Jablonka E, Lamb MJ (1995) *Epigenetic Inheritance and Evolution: the Lamarckian Dimension.* Oxford: Oxford University Press

Jackson CR (2003) Changes in community properties during microbial succession. *Oikos* 101: 444–448

Jagers op Akkerhuis GAJM (2008) Analysing hierarchy in the organization of biological and physical systems. *Biological Reviews* 83: 1–12

Jahn R (1998) C. G. Ehrenberg: The man and his botanical science. *Linnean* (special issue) 1: 15–28

Jain R, Rivera MC, Lake JA (1999) Horizontal gene transfer among genomes: the complexity hypothesis. *Proceedings of the National Academy of Sciences USA* 96: 3801–3806

James TY, Berbee ML (2011) No jacket required – new fungal lineage defies dress code. *BioEssays* 34: 94–102

Jannasch HW (1974) Steady state and the chemostat in ecology. *Limnology and Oceanography* 19: 716–720

—(1997) Small is powerful: recollections of a microbiologist and oceanographer. *Annual Review of Microbiology* 51: 1–45

Jannasch HW, Taylor CD (1984) Deep-sea microbiology. *Annual Review of Microbiology* 38: 487–514

Jarrell KF, Albers S-V (2012) The archeallum: an old motility structure with a new name. *Trends in Microbiology* 20: 307–312

Jeffery IB, Claesson MJ, O'Toole PW (2012) Categorization of the gut microbiota: enterotypes or gradients? *Nature Reviews Microbiology* 10: 591–592

Jenkins DG, Brescacin CR, Duxbury CV, Elliott JA, Evans JA, Grablow KR, Hillegass M, Lyon BN, Metzger GA, Olandese ML, et al. (2007) Does size matter for dispersal distance? *Global Ecology and Biogeography* 16: 415–425

Jessup CM, Kassen R, Forde SE, Kerr B, Buckling A, Rainey PB, Bohannan BJM (2004) Big questions, small worlds: microbial model systems in ecology. *Trends in Ecology and Evolution* 19: 189–197

Jessup CM, Forde SE, Bohannan BJM (2005) Microbial experimental systems in ecology. In Desharnais RA (ed), *Advances in Ecological Research*, Vol. 37 (pp. 273–307). London: Academic Press

Jogler C, Kube M, Schübbe S, Ullrich S, Teeling H, Bazylinski DA, Reinhardt R, Schüler D (2009) Comparative analysis of magnetosome gene clusters in magnetotactic bacteria provides further evidence for horizontal gene transfer. *Environmental Microbiology* 11: 1267–1277

Johnston RJ Jr, Desplan C (2010) Stochastic mechanisms of cell fate specification that yield random or robust outcomes. *Annual Review of Cell and Developmental Biology* 26: 689–719

Jones MDM, Forn I, Gadelha C, Egan MJ, Bass D, Massana R, Richards TA (2011) Discovery of novel intermediate forms redefines the fungal tree of life. *Nature* 474: 200–203

Jørgensen BB (2012) Shrinking majority of the deep biosphere. *Proceedings of the National Academy of Sciences USA* 109: 15976–15977

Justus J, Colyvan M, Regan H, Maguire L (2009) Buying into conservation: intrinsic versus instrumental value. *Trends in Ecology and Evolution* 24: 187–191

Kallmeyer J, Pockalny R, Adhikari RR, Smith DC, D'Hondt S (2012) Global distribution of microbial abundance and biomass in subseafloor sediment. *Proceedings of the National Academy of Sciences* 109: 16213–16216

Karr JB, Sanghvi JC, Macklin DN, Gutschow MV, Jacobs JM, Bolival B Jr, Assad-Garcia N, Glass JI, Covert MW (2012) A whole-cell computational model predicts phenotype from genotype. *Cell* 150: 389–401

Kashefi K, Lovley DR (2003) Extending the upper temperature limit for life. *Science* 301: 934

Kasting JF, Siefert JL (2002) Life and the evolution of earth's atmosphere. *Science* 296: 1066–1068

Kawecki TJ, Lenski RE, Ebert D, Hollis B, Olivieri I, Whitlock MC (2012) Experimental evolution. *Trends in Ecology and Evolution* 27: 547–560

Kay LE (2000) *Who Wrote the Book of Life? A History of the Genetic Code.* Stanford, CA: Stanford University Press

Keeling PJ, Palmer JD (2008) Horizontal gene transfer in eukaryotic evolution. *Nature Reviews Genetics* 9: 605–618

Keim CN, Martins JL, Abreu F, Rosado AS, de Barros HL, Borojevic R, Lins U, Farina M (2004) Multicellular life cycle of magnetotactic bacteria. *FEMS Microbiology Letters* 240: 203–208

Keller L, Surette MG (2006) Communication in bacteria: an ecological and evolutionary perspective. *Nature Reviews Microbiology* 4: 249–258

Kellert SH, Longino HE, Waters CK (2006) Introduction: the pluralist stance. In Kellert SH, Longino HE, Waters CK (eds), *Scientific Pluralism* (pp. vii–xxix). Minneapolis, MN: University of Minnesota Press

Kembel SW, Wu M, Eisen JA, Green JL (2012) Incorporating 16S gene copy number information improves estimates of microbial diversity and abundance. *PLoS Computational Biology* 8(10):e1002743, doi:10371/journal pcbi.1002743

Kiers ET, Duhamel M, Beesetty Y, Mensah JA, Franken O, et al. (2011) Reciprocal rewards stabilize cooperation in the mycorrhizal symbiosis. *Science* 333: 880–882

Kint CI, Verstraeten N, Fauvart M, Michiels J (2012) New-found fundamentals of bacterial persistence. *Trends in Microbiology* 20: 577–585

Kipling R (1902) *Just-So Stories.* London: MacMillan. Available at: http://www.boop.org/jan/justso/

Kirkpatrick CL, Viollier PH (2012) Decoding *Caulobacter* development. *FEMS Microbiology Reviews* 36: 193–205

Kirschner M, Gerhart J (1998) Evolvability. *Proceedings of the National Academy of Sciences USA* 95: 8420–8427

Kloesges T, Popa O, Martin W, Dagan T (2011) Networks of gene sharing among 329 proteobacterial genomes reveal differences in lateral gene transfer frequency at different phylogenetic depths. *Molecular Biology and Evolution* 28: 1057–1074

Kluyver AJ, Donker HJL (1924) The unity in the chemistry of fermentative sugar dissimilation processes in microbes. *Proceedings of the Koninklijke Nederlandse Akademie van Wetenschappen* 28 (3): 297–313. Available online at: www.dwc.knaw.nl/english/academy/digital-library

Kluyver AJ, van Niel CB (1956) *The Microbe's Contribution to Biology.* Cambridge, MA: Harvard University Press

—(1959 [1924]) Unity and diversity in the metabolism of microorganisms. In Kamp AF, La Rivière JWM, Verhoeven W (eds), *Albert Jan Kluyver, His Life and Work* (pp. 247–261). Amsterdam: North Holland. Available online at: www.archive.org/details/albertjankluyverookluy

Knoll AH (2003) The geological consequences of evolution. *Geobiology* 1: 3–14

—(2011) The multiple origins of complex multicellularity. *Annual Review of Earth and Planetary Sciences* 39: 217–239

Knoll AH, Javaux EJ, Hewitt D, Cohen D (2006) Eukaryotic organisms in Proterozoic oceans. *Philosophical Transactions of the Royal Society B* 361: 1023–1038

Koeppel A, Perry EB, Sikorski J, Krizanc D, Warner A, Ward DM, Rooney AP, Brambilla E, Connor N, Ratcliff RM, et al. (2008) Identifying the fundamental units of bacterial diversity: a paradigm shift to incorporate ecology into bacterial systematics. *Proceedings of the National Academy of Sciences USA* 105: 2504–2509

Koga R, Meng X-Y, Tsuchida T, Fukatsu T (2012) Cellular mechanism for selective vertical transmission of an obligate insect symbiont at the bacteriocyte-embryo interface. *Proceedings of the National Academy of Sciences USA* 109: E1230–1237

Kolber Z (2007) Energy cycle in the ocean: powering the microbial world. *Oceanography* 20: 79–88

Komeili A (2012) Molecular mechanisms of compartmentalization and biomineralization in magnetotactic bacteria. *FEMS Microbiology Reviews* 36: 232–255

Konopka A (2006) Microbial ecology: searching for principles. *Microbe* 1: 175–179

Koonin EV (2009) Darwinian evolution in the light of genomics. *Nucleic Acids Research* 37: 1011–1034

Koonin EV, Senkevich TG, Dolja VV (2006) The ancient virus world and evolution of cells. *Biology Direct* 1:29, doi:10.1186/1745-6150-1-29

Koonin EV, Wolf YI (2009) Is evolution Darwinian or/and Lamarckian? *Biology Direct* 4:42, doi:10.1186/1745-6150-4.42

Kopp RE, Kirschvink JL (2008) The identification and biogeochemical interpretation of fossil magnetotactic bacteria. *Earth-Science Reviews* 86: 42–61

Koren O, Goodrich JK, Cullender TC, Spor A, Laitinen K, Bäckhed HK, Gonzalez A, Werner JJ, Angenent LT, Knight R, et al. (2012) Host remodeling of the gut microbiome and metabolic changes during pregnancy. *Cell* 150: 470–480

Koschwanez JH, Foster KR, Murray (2011) Sucrose utilization in budding yeast as a model for the origin of undifferentiated multicellularity. *PLoS Biology* 9(8): e1001122

Koshland DE Jr (1980) *Bacterial Chemotaxis as a Model Behavioral System.* NY: Raven

Kuenen JG (2008) Anammox bacteria: from discovery to application. *Nature Reviews Microbiology* 6: 320–326

Lake JA, Henderson E, Oakes M, Clark MW (1984) Eocytes: a new ribosome structure indicates a kingdom with a close relationship to eukaryotes. *Proceedings of the National Academy of Sciences USA* 81: 3786–3790

Lan R, Reeves PR (2001) When does a clone deserve a name? A perspective on bacterial species based on population genetics. *Trends in Microbiology* 9: 419–424

Lane N, Martin W (2010) The energetics of genome complexity. *Nature* 467: 929–934

Lankau EW, Hong P-Y, Mackie RI (2012) Ecological drift and local exposures drive enteric bacterial community differences within species of Galápagos iguanas. *Molecular Ecology* 21: 1779–1788

Latty T, Beekman M (2010) Irrational decision-making in an amoeboid organism: transitivity and context-dependent preferences. *Proceedings of the Royal Society of London B* 278: 307–312

Lau JA, Lennon JT (2012) Rapid responses of soil microorganisms improve plant fitness in novel environments. *Proceedings of the National Academy of Sciences USA* 109: 14058–14062

Lawrence JG, Retchless AC (2010) The myth of bacterial species and speciation. *Biology and Philosophy* 25: 569–588

Lawrence JG, Hatfull GF, Hendrix RW (2002) Imbroglios of viral taxonomy: genetic exchange and failings of phenetic approaches. *Journal of Bacteriology* 184: 4891–4905

Lederberg J (1948) Problems in microbial genetics. *Heredity* 2: 145–198

—(1987) Genetic recombination in bacteria: a discovery account. *Annual Review of Genetics* 21: 23–46

Lederberg J, Tatum EL (1946) Gene recombination in *Escherichia coli. Nature* 158: 558

Lederberg J, Lederberg EM (1952) Replica plating and indirect selection of mutants. *Journal of Bacteriology* 63: 399–406

Leimar O, Hammerstein P (2010) Cooperation for direct fitness benefits. *Philosophical Transactions of the Royal Society London B* 365: 2619–2626

Lennon JT, Jones SE (2011) Microbial seed banks: the ecological and evolutionary implications of dormancy. *Nature Reviews Microbiology* 9: 119–130

Lenski RE, Mittler JE (1993) Directed mutation (response to Cairns). *Science* 260: 1222–1224

Lenski RE, Travisano M (1994) Dynamics of adaptation and diversification: a 10,000-generation experiment with bacterial populations. *Proceedings of the National Academy of Sciences USA* 91: 6808–6814

Lenski RE, Rose MR, Simpson SC, Tadler SC (1991) Long-term experimental evolution in *Escherichia coli*. 1. Adaptation and divergence during 2,000 generations. *American Naturalist* 138: 1315–1341

Lenton TM, Schnellnhuber HJ, Szathmáry E (2004) Climbing the co-evolution ladder. *Nature* 431: 913

Letelier J-C, Cárdenas ML, Cornish-Bowden A (2011) From *L'Homme Machine* to metabolic closure: steps towards understanding life. *Journal of Theoretical Biology* 286: 100–113

Levin BR (1972) Coexistence of two asexual strains on a single resource. *Science* 175: 1272–1274

—(1981) Periodic selection, infectious gene exchange and the genetic structure of *E. coli* populations. *Genetics* 99: 1–23

—(1988) Frequency-dependent selection in bacterial populations. *Philosophical Transactions of the Royal Society London B* 319: 459–472

—(2011) Population geneticists discover bacteria and their genetic/molecular epidemiology. In Walk ST and Fang PCH (eds), *Population Genetics of Bacteria: A Tribute to Thomas S. Whittam* 9 (pp. 7–13). Washington DC: ASM Press.

Levin BR, Bergstrom CT (2000) Bacteria are different: observations, interpretations, speculations and opinions about the mechanisms of adaptive evolution in prokaryotes. *Proceedings of the National Academy of Sciences USA* 97: 6981–6985

Levin BR, Rozen DE (2006) Non-inherited antibiotic resistance. *Nature Reviews Microbiology* 4: 556–562

Levy A, Currie A (2014) Model organisms are not (theoretical) models. *British Journal for Philosophy of Science* http://bjps.oxfordjournals.org/content/early/2014/03/26/bjps.axto55.short?rss=1

Lewin R (1981) Lamarck will not lie down. *Science* 213: 316–321

Ley RE, Turnbaugh PJ, Klein S, Gordon JI (2006) Human gut microbes associated with obesity. *Nature* 444: 1022–1023

Li K, Bihan M, Yooseph S, Methé BA (2012) Analyses of the microbial diversity across the human microbiome. *PLoS One* 7(6):e32118

Linares JF, Gustafsson I, Baquero F, Martinez JL (2006) Antibiotics as intermicrobial signaling agents instead of weapons. *Proceedings of the National Academy of Sciences USA* 103: 19484–19489

Lindström ES, Langenheder S (2011) Local and regional factors influencing bacterial community assembly. *Environmental Microbiology Reports* 4: 1–9

Little AEF, Robinson CJ, Peterson SB, Raffa KF, Handelsman J (2008) Rules of engagement: interspecies interactions that regulate microbial communities. *Annual Review of Microbiology* 62: 375–401

Lomolino MV (2000a) Ecology's most general, yet protean pattern: the species-area relationship. *Journal of Biogeography* 27: 17–26

—(2000b) A call for a new paradigm of island biogeography. *Global Ecology and Biogeography* 9: 1–6

Looft T, Johnson TA, Allen HK, Bayles DO, Alt DP, Stedtfeld RD, Sul WJ, Stedtfeld TM, Chai B, Cole JR (2012) In-feed antibiotic effects on the swine intestinal microbiome. *Proceedings of the National Academy of Sciences USA* 109: 1691–1696

Lopez D, Vlamakis H, Kolter R (2008) Generation of multiple cell types in *Bacillus subtilis*. *FEMS Microbiology Reviews* 33: 152–163

López-García P, Moreira D (2012) Viruses in biology. *Evolution: Education and Outreach* 5: 389–398

Loreau M (2010) Linking biodiversity and ecosystems: towards a unifying ecological theory. *Philosophical Transactions of the Royal Society London B* 365: 49–60

Love AC (2012) Hierarchy, causation and explanation: ubiquity, locality and pluralism. *Interface Focus* 2: 115–125

Love AC, Travisano M (2013) Microbes modeling ontogeny. *Biology and Philosophy* 28: 161–188

Lozupone CA, Knight R (2008) Species divergence and the measurement of microbial diversity. *FEMS Microbiology Reviews* 32: 557–578

Lozupone CA, Stombaugh JI, Gordon JI, Jansson JK, Knight R (2012) Diversity, stability and resilience of the human gut microbiota. *Nature* 489: 220–230

Lukjancenko O, Wassenaar TM, Ussery DW (2010) Comparison of 61 sequenced *Escherichia coli* genomes. *Microbial Ecology* 60: 708–720

Luria SE, Delbrück M (1943) Mutations of bacteria from virus sensitivity to virus resistance. *Genetics* 28: 491–511

Lwoff A (1957) The concept of virus. *Journal of General Microbiology* 17: 239–253

Lynch M (2007) The frailty of adaptive hypotheses for the origins of organismal complexity. *Proceedings of the National Academy of Sciences USA* 104: 8597–8604

MacArthur RH, Wilson EO (1967) *The Theory of Island Biogeography*. Princeton, NJ: Princeton University Press

Maclaurin J, Sterelny K (2008) *What is Biodiversity?* Chicago: University of Chicago Press

Madigan MT, Martinko JM, Dunlap PV, Clark DP (2008) *Brock Biology of Microorganisms* (12th ed). San Francisco, CA: Pearson/Benjamin Cummings

Mallet J (2008) Hybridization, ecological races and the nature of species: empirical evidence for the ease of speciation. *Philosophical Transactions of the Royal Society London B* 363: 2971–2986

Manichanh C, Reeder J, Gibert P, Varela E, Llopis M, Antolin M, Guigo R, Knight R, Guarner F (2010) Reshaping the gut microbiome with bacterial transplantation and antibiotic intake. *Genome Research* 20: 1411–1419

Margulis L (1971) Whittaker's five kingdoms of organisms: minor revisions suggested by considerations of the origin of mitosis. *Evolution* 25: 242–245

Martin GW (1955) Are fungi plants? *Mycologia* 47: 619–778

Martin W (2007) Eukaryote and mitochondrial origins: two sides of the same coin and too much ado about oxygen. In Falkowski PG, Knoll AH (eds), *Evolution of Primary Producers in the Sea* (pp. 55–73). Amsterdam: Elsevier

Martin W, Koonin EV (2006) A positive definition of prokaryotes. *Nature* 442: 868

Martinez JL (2009) The role of natural environments in the evolution of resistance traits in pathogenic bacteria. *Proceedings of the Royal Society London B* 276: 2521–2530

Martiny JBH, Bohannan BJM, Brown JH, Colwell RK, Fuhrman JA, Green JL, Horner-Devine MC, Kane M, Krumins JA, Kuske CR, et al. (2006) Microbial biogeography: putting microorganisms on the map. *Nature Reviews Microbiology* 4: 102–112

Materna AC, Firedman J, Bauer C, David C, Chen S, Huang IB, Gillens A, Clarke SA, Polz MF, Alm EJ (2012) Shape and evolution of the fundamental niche in marine *Vibrio*. *ISME Journal* 6: 2168–2177

Maynard Smith J (1995) Do bacteria have population genetics? In Baumberg S., Young JPW, Wellington EMH, Saunders JR (eds), *Population Genetics of Bacteria* (pp. 1–12). Cambridge, UK: Cambridge University Press.

Maynard Smith J, Szathmáry E (1997) *The Major Transitions In Evolution*. Oxford: Oxford University Press

Maynard Smith J, Smith NH, O'Rourke M, Spratt BG (1993) How clonal are bacteria? *Proceedings of the National Academy of Sciences USA* 90: 4384–4388

Mayr E (1942) *Systematics and the Origin of Species From the Viewpoint of a Zoologist*. NY: Columbia University Press

—(1953) Concepts of classification and nomenclature in higher organisms and microorganisms. *Annals of the New York Academy of Sciences* 56: 391–397

—(1963) *Animal Species and Evolution*. Cambridge, MA: Harvard University Press

—(1987) The ontological status of species: scientific progress and philosophical terminology. *Biology and Philosophy* 2: 145–166

—(1998) Two empires or three? *Proceedings of the National Academy of Sciences USA* 95: 9720–9723

—(2001) *What Evolution Is*. Basic Books, NY

—(2004) The evolution of Ernst: interview with Ernst Mayr. *Scientific American*, July 6th. http://www.scientificamerican.com/article.cfm?id=the-evolution-of-ernst-in

Mazumdar PMH (1995) *Species and Specificity: An Interpretation of the History of Immunology*. Cambridge: Cambridge University Press

McCormick JR, Flärdh K (2012) Signals and regulators that govern *Streptomyces* development. *FEMS Microbiology Reviews* 36: 206–231

McFall-Ngai M, Hadfield MG, Bosch TCG, Carey HV, Domazet-Lošo T, Douglas AE, Dubilier N, Eberl G, Fukami T, Gilbert SF, et al. (2013) Animals in a bacterial world, a new imperative for the life sciences. *Proceedings of the National Academy of Sciences USA* 110: 3229–3236

McGill BJ, Maurer BA, Weiser MD (2006) Empirical evaluation of neutral theory. *Ecology* 87: 1411–1423

McGuinness KA (1984) Equations and explanations in the study of species-area curves. *Biological Reviews* 59: 423–440

McInerney JO, Martin WF, Koonin EV, Allen JF, Galperin MY, Lane N, Archibald JM, Embley TM (2011) Planctomycetes and eukaryotes: a case of analogy not homology. *BioEssays* 33: 810–817

McInerney JO, Pisani D, Bapteste E, O'Connell MJ (2011) The public goods hypothesis for the evolution of life on Earth. *Biology Direct* 6:41, doi:10.1186/1745-6150-6-41

McShea DW, Simpson C (2011) The miscellaneous transitions in evolution. In Calcott B, Sterelny K (eds), *The Major Transitions in Evolution Revisited* (pp. 17–33) Cambridge, MA: MIT Press

Mellon RR (1925) Studies in microbic heredity: 1. Observations on a primitive form of sexuality (zygospore formation) in the colon-typhoid group. *Journal of Bacteriology* 10: 481–501 (plus 5 plates)

Mendelsohn JA (2002) 'Like all that lives': Biology, medicine and bacteria in the age of Pasteur and Koch. *History and Philosophy of the Life Sciences* 24: 3–36

Metcalf JA, Bordenstein SR (2012) The complexity of virus systems: the case of endosymbionts. *Current Opinion in Microbiology* 15: 546–552

Michael CA, Gillings MR, Holmes AJ, Hughes L, Andrew NR, Holley MP, Stokes HW (2004) Mobile gene cassettes: a fundamental resource for bacterial evolution. *American Naturalist* 164: 1–12

Millikan RG (1989) Biosemantics. *Journal of Philosophy* 86: 281–297

—(2007) An input condition for teleosemantics? Reply to Shea (and Godfrey-Smith). *Philosophy and Phenomenological Research* 75: 436–455

Mitri S, Xavier JB, Foster KR (2011) Social evolution in multispecies biofilms. *Proceedings of the National Academy of Sciences USA* 108: 10839–10846

Monds RD, O'Toole GA (2009) The developmental model of microbial biofilms: ten years of a paradigm up for review. *Trends in Microbiology* 17: 73–87

Mongodin EF, Nelson KE, Daugherty S, DeBoy RT, Wister J, Khouri H, Weidman J, Walsh DA, Papke RT, Perez GS, et al. (2005) The genome of *Salinibacter ruber*: convergence and gene exchange among hyperhalophilic bacteria and archaea. *Proceedings of the National Academy of Sciences USA* 102: 18147–18152

Monod J, Jacob F (1961) General conclusions: teleonomic mechanisms in cellular metabolism, growth, and differentiation. *Cold Spring Harbor Symposia on Quantitative Biology* 26: 389–401

Moore RC (1954) Kingdom of organisms named Protista. *Journal of Paleontology* 28: 588–598

Mora C, Tittensor DP, Adl S, Simpson AGB, Worm B (2011) How many species are there on Earth and in the ocean? *PLoS Biology* 9(8):e1001127

Moran NA, McCutcheon JP, Nakabachi A (2008) Genomics and the evolution of heritable bacterial symbionts. *Annual Review of Genetics* 42: 165–190

Moreira D, Brochier-Armanet C (2008) Giant viruses, giant chimeras: the multiple evolutionary histories of Mimivirus genes. *BMC Evolutionary Biology* 8:12, doi:10.1186/1471-2148-8/12

Moreira D, López-García P (2009) Ten reasons to exclude viruses from the tree of life. *Nature Reviews Microbiology* 7: 306–311

Morris BEL, Henneberger R, Huber H, Moissl-Eichinger C (2013) Microbial syntrophy: interaction for the common good. *FEMS Microbiology Reviews* 37: 384–406

Moya A, Peretó J, Gil R, Latorre A (2008) Learning how to live together: genomic insights into prokaryote-animal symbioses. *Nature Reviews Genetics* 9: 218–229

Müller-Hill B (1996) *The Lac Operon: A Short History of a Genetic Paradigm.* Berlin: De Gruyter

Murray RGE, Holt JG (2005) The history of Bergey's Manual. In *Bergey's Manual of Systematic Bacteriology, Volume Two: The Proteobacteria* (pp. 1–14). NY: Springer

Nachtomy O, Shavit A, Smith J (2002) Leibnizian organisms, nested individuals, and units of selection. *Theory in Biosciences* 121: 205–230

Nakamura Y, Itoh T, Matsuda H, Gojobori T (2004) Biased biological functions of horizontally transferred genes in prokaryotic genomes. *Nature Genetics* 36: 760–766

Naumann M, Schüßler A, Bonfante P (2010) The obligate endobacteria of arbuscular mycorrhizal fungi are ancient heritable components related to the *Mollicutes*. *ISME Journal* 4: 862–871

Neander K (2012) Teleological theories of mental content. In Zalta EN (ed), *Stanford Encyclopedia of Philosophy*, http://plato.stanford.edu/entries/content-teleological/

Nelson D (2004) Phage taxonomy: we agree to disagree. *Journal of Bacteriology* 186: 7029–7031

Nelson-Sathi S, Dagan T, Landan G, Janssen A, Steel M, McInerney JO, Deppenmeier U, Martin WF (2012) Acquisition of 1,000 eubacterial genes physiologically transformed an methanogen at the origin of Haloarchaea. *Proceedings of the National Academy of Sciences USA* 109: 20537–20542

NIH (2012) NIH Human Microbiome Project defines normal bacterial makeup of the body, June 13th. www.nih.gov/news/health/jun2012/nhgri-13.htm

Niklas KJ, Newman SA (2013) The origins of multicellular organisms. *Evolution and Development* 15: 41–52

Nikoh N, Tanaka K, Shibata F, Kondo N, Hizume M, Shimada M, Fukatsu T (2008) *Wolbachia* genome integrated in an insect chromosome: evolution and fate of laterally transferred endosymbiont genes. *Genome Research* 18: 272–280

Nishida K, Silver PA (2012) Induction of biogenic magnetization and redox control by a component of the target of rapamycin complex I signaling pathway. *PLoS Biology* 10(2): e1001269

O'Malley MA (2007) The nineteenth century roots of 'Everything is everywhere'. *Nature Reviews Microbiology* 5: 647–651

—(2009) What *did* Darwin say about microbes, and how did microbiology respond? *Trends in Microbiology* 17: 341–347

—(2010a) Ernst Mayr, the Tree of Life, and philosophy of biology. *Biology and Philosophy* 25: 529–552

—(2010b) The first eukaryote cell: an unfinished history of contestation. *Studies in History and Philosophy of Biological and Biomedical Sciences* 41: 212–224

O'Malley MA, Dupré J (2005) Fundamental issues in systems biology. *BioEssays* 27: 1270–1276

—(2007) Size doesn't matter: towards a more inclusive philosophy of biology. *Biology and Philosophy* 22: 155–191

O'Malley MA, Koonin EV (2011) How stands the Tree of Life a century and a half after The Origin? *Biology Direct* 6:32, doi:10.1186/1745-6150-6-32

O'Malley MA, Simpson AGB, Roger AJ (2013) The other eukaryotes in light of evolutionary protistology. *Biology and Philosophy* 28: 299–330

O'Malley MA, Powell A, Davies JF, Calvert J (2007) Knowledge-making distinctions in synthetic biology. *BioEssays* 30: 57–65

O'Toole G, Kaplan HB, Kolter R (2000) Biofilm formation as microbial development. *Annual Review of Microbiology* 54: 49–79

Ochman H, Worobey M, Kuo C-H, Njdango J-BN, Peeters M, Hahn BH, Hugenholtz P (2010) Evolutionary relationships of wild hominids recapitulated by gut microbial communities. *PLoS Biology* 8(11):e1000546

Odenbaugh J (2006) Struggling with the science of ecology. *Biology and Philosophy* 21: 395–409

—(2007) Seeing the forest and the trees: realism about communities and ecosystems. *Philosophy of Science* 74: 628–641

Odum EP (1971) *Fundamentals of Ecology* (3rd ed). Philadelphia: Saunders

Okasha S (2011) Biological ontology and hierarchical organization: a defense of rank freedom. In Calcott B, Sterelny K (eds), *Major Transitions in Evolution Revisited* (pp. 53–63). Cambridge MA: MIT Press

Olson JM, Blankenship RE (2004) Thinking about the evolution of photosynthesis. *Photosynthesis Research* 80: 373–386

Omer S, Kovacs A, Mazor Y, Gophna U (2010) Integration of a foreign gene into a native complex does not impair fitness in an experimental model of lateral gene transfer. *Molecular Biology and Evolution* 27: 2441–2445.

Oppenheim AB, Kobiler O, Stavans J, Court DL, Adhya S (2005) Switches in bacteriophage lambda development. *Annual Review of Genetics* 39: 409–429

Oren A (2004) Prokaryote diversity and taxonomy: current status and future challenges. *Philosophical Transactions of the Royal Society London B* 359: 623–638

—(2011) The halophilic world of Lourens Bass Becking. In Ventosa A, Oren A, Ma Y (eds), *Halophiles and Hypersaline Environments: Current Research and Future Trends* (pp. 9–25). Berlin: Springer-Verlag

Pace NR (1997) A molecular view of microbial diversity and the biosphere. *Science* 276: 734–740

—(2006) Time for a change. *Nature* 441: 289

—(2009) Rebuttal: the modern concept of the prokaryote. *Journal of Bacteriology* 191: 2006–2007

Pace NR, Sapp J, Goldenfeld N (2012) Phylogeny and beyond: scientific, historical, and conceptual significance of the first tree of life. *Proceedings of the National Academy of Sciences USA* 109: 1011–1018

Papke RT, Ramsing NB, Bateson MM, Ward DM (2003) Geographical isolation in hot spring cyanobacteria. *Environmental Microbiology* 5: 650–659

Parfrey LW, Lahr DJG (2013) Multicellularity arose several times in the evolution of eukaryotes. *BioEssays* doi:10.1002/bies.201200143

Peiffer JA, Spor A, Koren O, Jin Z, Tringe SG, Dandl JL, Buckler ES, Ley RE (2013) Diversity and heritability of the maize rhizosphere microbiome under field conditions. *Proceedings of the National Academy of Sciences USA* 110: 6548–6553

Pepper JW, Rosenfeld S (2012) The emerging medical ecology of the human gut microbiome. *Trends in Ecology and Evolution* 27: 381–384

Philippe H, Brinkmann H, Lavrov DV, Littlewood DTJ, Manuel M, Wörheide G, Barain D (2011) Resolving difficult phylogenetic questions: why more sequences are not enough. *PLoS Biology* 9(3): e1000602

Pigliucci M (2008) Is evolvability evolvable? *Nature Reviews Genetics* 9: 75–82

Pomeroy LR (1974) The ocean's food web, a changing paradigm. *BioScience* 24: 499–504

Pomeroy LR, Williams PJ leB, Azam F, Hobbie JE (2007) The microbial loop. *Oceanography* 20: 28–33

Poole AM, Phillips MJ, Penny D (2003) Prokaryote and eukaryote evolvability. *BioSystems* 69: 163–185

Potochnik A, McGill B (2012) The limitations of hierarchical organization. *Philosophy of Science* 79: 120–140

Pradeu T (2012) *The Limits of Self: Immunology and Biological Identity* (Vitanza E, transl). Oxford: Oxford University Press

Prosser JI (2010) Replicate or lie. *Environmental Microbiology* 12: 1806–1810

Prosser JI, Bohannan BJM, Curtis TP, Ellis RJ, Firestone MK, Freckelton RP, Green JL, Green LE, Kilham K, Lennon JJ, et al. (2007) The role of ecological theory in microbial ecology. *Nature Reviews Microbiology* 5: 384–392

Psenner R, Alfreider A, Schwartz A (2008) Aquatic microbial ecology: water desert, microcosm, ecosystem. What's next? *International Review of Hydrobiology* 93: 606–623

Puigbò P, Wolf YI, Koonin EV (2009) The tree and net components of prokaryote evolution. *Genome Biology and Evolution* 2: 745–756

Purnick PEM, Weiss R (2010) The second wave of synthetic biology: from modules to systems. *Nature Reviews Molecular Cell Biology* 10: 410–422

Qin J, Li R, Raes J, Arumugam M, Burgdorf KS, Manichanh C, Nielsen T, Pons N, Levenez F, Yamada T, et al. (2010) A human gut microbial gene catalogue established by metagenomic sequencing. *Nature* 464: 59–65

Queller DC (1997) Cooperators since life began. *Quarterly Review of Biology* 72: 184–188

Queller DC, Strassmann JE (2009) Beyond society: the evolution of organismality. *Philosophical Transactions of the Royal Society London B* 364: 3143–3155

Quispel A (1998) Lourens G. M. Baas Becking (1895–1963), inspirator for many (micro)biologists. *International Microbiology* 1: 69–72

Raes J, Letunic I, Yamada T, Jense LJ, Bork P (2011) Toward molecular trait-based ecology through integration of biogeochemical, geographical and metagenomic data. *Molecular Systems Biology* 7:473, doi:10.1038/msb.2011.6

Ragan MA, Beiko RG (2009) Lateral genetic transfer: open issues. *Philosophical Transactions of the Royal Society London B* 364: 2241–2251

Ragon M, Fontaine MC, Moreira D, López-García P (2012) Different biogeographic patterns of prokaryotes and microbial eukaryotes in epilithic biofilms. *Molecular Ecology* 21: 3852–3868

Rainey PB, Buckling A, Kassen R, Travisano M (2000) The emergence and maintenance of diversity: insights from experimental bacterial populations. *Trends in Ecology and Evolution* 15: 243–247

Rainey PB, Beaumont HJE, Ferguson GC, Gallie J, Kost C, Libby, Zhang XX (2011) The evolutionary emergence of stochastic phenotype switching in bacteria. *Microbial Cell Factories* 10 (Suppl 1): S14, doi:10.1186/1475-2859-10-S1-S14

Rainey PB, Kerr B (2010) Cheats as first propagules: a new hypothesis for the evolution of individuality during the transition from single cells to multicellularity. *BioEssays* 32: 872–880

Rankin DJ, Bichsel M, Wagner A (2010) Mobile DNA can drive lineage extinction in prokaryotic populations. *Journal of Evolutionary Biology* 23: 2422–2431

Raoult D, Forterre P (2008) Redefining viruses: lessons from Mimivirus. *Nature Reviews Microbiology* 6: 315–319

Raoult D, Audic S, Robert C, Abergel C, Renesto P, Ogata H, La Scola B, Suzan M, Claverie -JM (2004) The 1.2-megabase genome sequence of mimivirus. *Science* 306: 1344–1350

Ratcliff MJ (2009) *The Quest for the Invisible: Microscopy in the Enlightenment.* Farnham, Surrey: Ashgate

Ratcliff WC, Denison RF, Borrello M, Travisano M (2012) Experimental evolution of multicellularity. *Proceedings of the National Academy of Sciences USA* 109: 1595–1600

Raymond J, Segrè D (2006) The effect of oxygen on biochemical networks and the evolution of complex life. *Science* 311: 1764–1767

Reche I, Pulido-Villena E, Morales-Baquero R, Casamayor EO (2005) Does ecosystem size determine aquatic bacterial richness? *Ecology* 86: 1715–1722

Redfield R (1993) Genes for breakfast: the have-your-cake-and-eat-it-too of bacterial transformation. *Journal of Heredity* 84: 400–404

—(2001) Do bacteria have sex? *Nature Reviews Genetics* 2: 634–639

Reed CD, Christopher KA, Huber JA, Dick GJ (2014) Gene-centric approach to integrating environmental genomics and biogeochemical models. *Proceedings of the National Academy of Sciences USA* 111: 1879-1884

Reid CR, Latty T, Dussutour A, Beekman M (2012) Slime mold uses and externalized spatial 'memory' to navigate in complex environments. *Proceedings of the National Academy of Sciences USA* 109: 17490–17494

Relman DA (2013) Undernutrition – looking within for answers. *Science* 339: 530–532

Renault S, Stasiak K, Federici B, Bigot Y (2005) Commensal and mutualistic relationships of reoviruses with their parasitoid wasp hosts. *Journal of Insect Physiology* 51: 137–148

Replansky T, Koufopanou V, Greig D, Bell G (2008) *Saccharomyces sensu stricto* as a model system for evolution and ecology. *Trends in Ecology and Evolution* 23: 494–501

Reyes-Prieto A, Weber APM, Bhattacharya D (2007) The origin and establishment of the plastid in algae and plants. *Annual Review of Genetics* 41: 147–168

Richards TA (2011) Genome evolution: horizontal movements in the fungi. *Current Biology* 21: R166–R168

Richards TA, Talbot NJ (2013) Horizontal gene transfer in osmotrophs: playing with public goods. *Nature Reviews Microbiology* 11: 720–727

Ricklefs RE (2011) Applying a regional community concept to forest birds of eastern North America. *Proceedings of the National Academy of Sciences USA* 108: 2300–2305

Rivera MC, Lake JA (2004) The ring of life provides evidence for a genome fusion origin of eukaryotes. *Nature* 431: 152–155

Robertson LA (2003) The Delft school of microbiology, from the nineteenth to the twenty-first century. *Advances in Applied Microbiology* 52: 357–388

Robinson CJ, Bohannan BJM, Young VB (2010) From structure to function: the ecology of host-associated microbial communities. *Microbiology and Molecular Biology Reviews* 74: 453–476

Rocha EPC (2003) An appraisal of the potential for illegitimate recombination in bacterial genomes and its consequences: from duplications to genome reduction. *Genome Research* 13: 1123–1132

Rogers GB, Carroll MP, Bruce KD (2009) Studying bacterial infections through culture-independent approaches. *Journal of Medical Microbiology* 58: 1401–1418

Rohwer F, Thurber RV (2009) Viruses manipulate the marine environment. *Nature* 459: 207–212

Rohwer F, Barott K (2013) Viral information. *Biology and Philosophy* 28: 283–297

Roossinck M (2011) The good viruses: viral mutualistic symbioses. *Nature Reviews Microbiology* 9: 99–108

Rose MR, Oakley TH (2007) The new biology: beyond the Modern Synthesis. *Biology Direct* 2:30, doi:10.1186/1745-6150-2-30

Rosenberg E, Sharon G, Zilber-Rosenberg I (2009) The hologenome theory of evolution contains Lamarckian aspects within a Darwinian framework. *Environmental Microbiology* 11: 2959–2962

Rosenberg SM (2001) Evolving responsively: adaptive mutation. *Nature Reviews Genetics* 2: 504–515

Rosenzweig RF, Sharp RR, Treves DS, Adams J (1994) Microbial evolution in a simple unstructured environment: genetic differentiation in *Escherichia coli*. *Genetics* 137: 903–917

Rosindell J, Hubbell SP, He F, Harmon LJ, Etienne RS (2012) The case for ecological neutral theory. *Trends in Ecology and Evolution* 27: 203–208

Roselló-Mora R, Amann R (2001) The species concept for prokaryotes. *FEMS Microbiology Reviews* 25: 39–67

Roth JR, Kugelberg E, Reams AB, Kofoid E, Andersson DI (2006) Origin of mutations under selection: the adaptive mutation controversy. *Annual Review of Microbiology* 60: 477–501

Rothschild LJ (1989) Protozoa, Protista, Protoctista: what's in a name? *Journal of the History of Biology* 22: 277–305

Sachs JL, Simms EL (2008) The origins of uncooperative rhizobia. *Oikos* 117: 961–966

Samuelson J (1865) On the development of certain infusoria. *Proceedings of the Royal Society London* 14: 546–547

Sapp J (2004) *Evolution by Association: A History of Symbiosis.* NY: Oxford University Press

—(2009) *The New Foundations of Evolution: On the Tree of Life.* NY: Oxford University Press

Saridaki A, Bourtzis K (2010) *Wolbachia*: more than just a bug in insects genitals. *Current Opinion in Microbiology* 13: 67–72

Savage DC (1977) Microbial ecology of the gastrointestinal tract. *Annual Review of Microbiology* 31: 107–133

Scanlan D (2001) Cyanobacteria: ecology, niche adaptation and genomics. *Microbiology Today* 28: 128–130

Schaap P (2007) Evolution of size and pattern in the social amoebas. *BioEssays* 29: 635–644

Schaechter M (2012a) The tyranny of phylogeny: an exhortation. *Small Things Considered*, October 25, http://schaechter.asmblog.org/schaechter/2012/10/the-tyranny-of-phylogeny-an-exhortation.html

—(2012b) Who would have thought it? *Small Things Considered*, September 6, http://schaechter.asmblog.org/schaechter/2012/09/who-would-have-thought-it.html

Schatz A (1993) The true story of the discovery of streptomycin. *Actinomycetes* 4: 27–39

Schatz A, Bugle E, Waksman SA (1944) Streptomycin, a substance exhibiting antibiotic activity against Gram-positive and Gram-negative bacteria. *Proceedings of the Society for Experimental Biology and Medicine NY* 55: 66–69

Schirrmeister BE, de Vos JM, Antonelli A, Bagheri HC (2013) Evolution of multicellularity coincided with increased diversification of cyanobacteria and the Great Oxidation Event. *Proceedings of the National Academy of Sciences USA* 110: 1791–1796

Schloissnig S, Arumugam M, Sunagawa S, Mitreva M, Tap J, Zhu A, Waller A, Mende DR, Kultima JR, Martin J, et al. (2013) Genomic variation landscape of the human gut microbiome. *Nature* 493: 45–50

Schluter J, Foster KR (2012) The evolution of mutualism in gut microbiota via host epithelial selection. *PLoS Biology* 10(11): e1001424

Schrödinger E (1944) *What is Life? The Physical Aspect of the Living Cell.* Cambridge: Cambridge University Press

Sessions AL, Doughty DM, Welander PV, Summons RE, Newman DK (2009) The continuing puzzle of the Great Oxidation Event. *Current Biology* 19: R567–R574

Shade A, Handelsman J (2012) Beyond the Venn diagram: the hunt for a core microbiome. *Environmental Microbiology* 14: 4–12

Shade A, Hogan CS, Klimowicz AK, Linske M, McManus PS, Handelsman J (2012) Culturing captures members of the rare soil biosphere. *Environmental Microbiology* 14: 2247–2252

Shapiro BJ, Friedman J, Cordero OX, Preheim SP, Timberlake SC, Szabó G, Polz MF, Alm EJ (2012) Population genomics of early events in the ecological differentiation of bacteria. *Science* 336: 48–51

Shapiro JA (1995) Genome organization, natural genetic engineering and adaptive mutation. *Trends in Genetics* 13: 98–104

—(1998) Thinking about bacterial populations as multicellular organisms. *Annual Review of Microbiology* 52: 81–104

Shapiro OH, Hatzenpichler R, Buckley DH, Zinder SH, Orphan VJ (2011) Multicellular photo-magnetotactic bacteria. *Environmental Microbiology Reports* 3: 233–238

Sharp NP, Agrawal AF (2012) Evidence for elevated mutation rates in low-quality genotypes. *Proceedings of the National Academy of Sciences USA* 109: 6142–6146

Shea N (2006) Millikan's contribution to materialist philosophy of mind. *Matière Première* 1: 127–156

—(2013) Naturalising representational content. *Philosophy Compass* 8: 496–509

Sibley CD, Surette MG (2011) The polymicrobial nature of airway infections in cystic fibrosis: Cangene Gold Medal Lecture. *Canadian Journal of Microbiology* 57: 69–77

Simberloff D (1976) Species turnover and equilibrium island biogeography. *Science* 194: 572–578

Simmons SL, Bazylinski DA, Edwards KJ (2006) South-seeking magnetotactic bacteria in the Northern Hemisphere. *Science* 311: 371–374

Simms EL, Taylor DL (2002) Partner choice in nitrogen-fixation mutualisms of legumes and rhizobia. *Integrative and Computational Biology* 42: 369–380

Simpson AGB, Roger AJ (2004) The real 'kingdoms' of eukaryotes. *Current Biology* 14: R693–R696

Singh S, Eldin C, Kowalczewska M, Raoult D (2013) Axenic culture of fastidious and intracellular bacteria. *Trends in Microbiology* 21: 92–99

Skippington E, Ragan MA (2012) Phylogeny rather than ecology or lifestyle biases the construction of *Escherichia coli-Shigella* genetic exchange communities. *Open Biology* 2: 120112, doi:10.1098/rsob.120112

Skyrms B (2010) *Signals: Evolution, Learning, and Information*. Oxford University Press

Slater MH (2014) Natural kindness. *British Journal for the Philosophy of Science*, doi:10.1093/bjps/axt033

Sloan P (2008) Evolution. In Zalta EN (ed), *Stanford Encyclopedia of Philosophy*, http://plato.stanford.edu/entries/evolution/

Sloan WT, Woodcock S, Lunn M, Head IM, Curtis TP (2007) Modeling taxa-abundance distributions in microbial communities using environmental sequence data. *Microbial Ecology* 53: 443–455

Smillie CS, Smith MB, Friedman J, Cordero OX, David LA, Alm EJ (2011) Ecology drives a global network of gene exchange connecting the human microbiome. *Nature* 480: 241–244

Smith MI, Yatsunenko T, Manary MJ, Trehan I, Mkakosya R, Cheng J, Kau AL, Rich SS, Concannon P, Mychaleckyj JC, et al. (2013) Gut microbiomes of Malawian twin pairs discordant for kwashiorkor. *Science* 339: 548–554

Smith T (1904) Some problems in the life history of pathogenic microorganisms. *Science* 20: 817–832

Smith VH (2007) Microbial diversity-productivity relationships in aquatic ecosystems. *FEMS Microbiology Ecology* 62: 181–186

Sniegowski PD, Lenski RE (1995) Mutation and adaptation: the directed mutation controversy in evolutionary perspective. *Annual Review of Ecology and Systematics* 26: 553–578

Sniegowski PD, Gerrish PJ, Johnson T, Shaver A (2000) The evolution or mutation rates: separating causes from consequences. *BioEssays* 22: 1057–1066

Sogin ML, Morrison HG, Huber JA, Welch DM, Huse SM, Neal PR, Arrieta JM, Herndl GJ (2006) Microbial biodiversity in the deep sea and the underexplored 'rare biosphere'. *Proceedings of the National Academy of Sciences USA* 103: 12115–12120

Somvanshi VS, Sloup RE, Crawford JM, Martin AR, Heidt AJ, Kim K-s, Clardy J, Ciche TA (2012) A single promoter inversion switches *Photorhabdus* between pathogenic and mutualistic states. *Science* 337: 88–93

Sonea S, Mathieu LG (2001) Evolution of the genomic systems of prokaryotes and its momentous consequences. *International Microbiology* 4: 67–71

Sonneborn TM (1965) Genetics and man's vision. *Proceedings of the American Philosophical Society* 109: 237–241

Sorokin Y (1971) Review of EJ Ferguson Wood's 'Microbiology of Oceans and Estuaries'. *Journal du Conseil International pour l'Éxploration de la Mer* 33: 515–516

Spath S (2004) Van Niel's course in general microbiology. *ASM News* 70: 359–363

Spratt BG, Hanage WP, Feil EJ (2001) The relative contributions of recombination and point mutation to the diversification of bacterial clones. *Current Opinion in Microbiology* 4: 602–606

Spring S, Amann R, Ludwig W, Schleifer K-H, van Gemerden H, Petersen N (1993) Dominating role of an unusual magnetotactic bacterium in the microaerobic zone of a freshwater sediment. *Applied and Environmental Microbiology* 59: 2397–2403

Stackebrandt E, Fredericksen W, Garrity GM, Grimont PAD, Kämpfer P, Maiden MCJ, Nesme X, Rosseló-Mora R, Sings J, Trüper HG, et al. (2002) Report of the ad hoc committee for the re-evaluation of the species definition in bacteriology. *International Journal of Systematic and Evolutionary Microbiology* 52: 1043–1047

Stahl DA, Lane DJ, Olsen GJ, Pace NR (1985) Characterization of a Yellowstone hot spring microbial community by 5S rRNA sequences. *Applied and Environmental Microbiology* 49: 1379–1384

Staley JT (1997) Biodiversity: are microbial species threatened? *Current Opinion in Biotechnology* 8: 340–345

Staley JT, Konopka A (1985) Measurement of in situ activities of non-photosynthetic microorganisms in aquatic and terrestrial habitats. *Annual Review of Microbiology* 39: 321–346

Stanier RY (1951) The life-work of a founder of bacteriology: a review of 'Microbiologie du Sol'. *Quarterly Review of Biology* 26: 35–37

Stanier RY, van Niel CB (1962) The concept of a bacterium. *Archive für Mikrobiologie* 42: 17–35

Stanier RY, Doudoroff M, Adelberg EA (1957) *The Microbial World*. Englewood Cliffs NJ: Prentice-Hall

Steele EJ (1981) *Somatic Selection and Adaptive Evolution: On the Inheritance of Acquired Characters* (2nd ed). Chicago: University of Chicago Press

Stegen JC, Lin X, Konopka AE, Frederickson JK (2012) Stochastic and deterministic assembly processes in subsurface microbial communities. *ISME Journal* 6: 1653–1664

Steinhaus EA (1960) The importance of environmental factors in the insect-microbe ecosystem. *Journal of Bacteriology* 24: 365–373

Sterelny K (1999) Bacteria at the high table. *Biology and Philosophy* 14: 459–470

—(2004) Symbiosis, evolvability and modularity. In Schlosser G, Wagner G (eds), *Modularity in Development and Evolution* (pp. 490–516). Chicago: University of Chicago Press

—(2006a) Local ecological communities. *Philosophy of Science* 73: 215–231

—(2006b) What is evolvability? In Matthen M, Stephens C (eds), *Handbook of the Philosophy of Science* (pp. 177–192). Amsterdam: Elsevier

Stone RW, ZoBell CE (1952) Bacterial aspects of the origin of petroleum. *Industrial and Engineering Chemistry* 44: 2564–2567

Straight PD, Kolter R (2009) Interspecies chemical communication in bacterial development. *Annual Review of Microbiology* 63: 99–118

Strassmann JE, Queller DC (2010) The social organism: congresses, parties, and committees. *Evolution* 64: 605–616

Summers WC (1991) From culture as organism to organism as cell: historical origins of bacterial genetics. *Journal of the History of Biology* 24: 171–190

Suttle CA (2007) Marine viruses – major players in the global ecosystem. *Nature Reviews Microbiology* 5: 801–812

Swenson W, Wilson DS, Elias R (2000) Artificial ecosystem selection. *Proceedings of the National Academy of Sciences USA* 97: 9110–9114

Szathmáry E, (2012) Transitions and social evolution: a review of *Principles of Social Evolution*. *Philosophy & Theory in Biology* 4: e302, doi:10.3998/ptb.6959004.0004.002

Szathmáry E, Maynard Smith J (1995) The major evolutionary transitions. *Nature* 374: 227–232

Szathmáry E, Fernando C (2011) Concluding remarks. In Calcott B, Sterelny K (eds), *The Major Transitions in Evolution Revisited* (pp. 301–310) Cambridge, MA: MIT Press

Szostak JW (2012) Attempts to define life do not help to understand the origin of life. *Journal of Biomolecular Structure and Dynamics* 29: 599–600

Tanouchi Y, Pai A, Buchler NE, You L (2012) Programming stress-induced altruistic death in engineered bacteria. *Molecular Systems Biology* 8: 626, doi:10.1038/msb.2012.57

Tansley AG (1935) The use and abuse of vegetational concepts and terms. *Ecology* 16: 284–307

Tatum EL (1959) A case history in biological research (Nobel lecture). *Science* 129: 1711–1715

Tenaillon O, Taddei F, Radman M, Matic I (2001) Second-order selection in bacterial evolution: selection acting on mutation and recombination rates in the course of adaptation. *Research in Microbiology* 152: 11–16

Tettelin H, Masignani V, Cieslewicz MJ, Donati C, Medini D, Ward NL, Angiuoli SV, Crabtree J, Jones Al, Durkin AS, et al. (2005) Genome analysis of multiple pathogenic isolates of *Streptococcus agalactiae*: implications for the microbial 'pan-genome'. *Proceedings of the National Academy of Sciences USA* 102: 13950–13955

Theunissen B (1996) The beginnings of the 'Delft tradition' revisited: Martinus W. Beijerinck and the genetics of microorganisms. *Journal of the History of Biology* 29: 197–228

Thomas CM, Nielsen KM (2005) Mechanisms of, and barriers to, horizontal gene transfer between bacteria. *Nature Reviews Microbiology* 3: 711–721

Tiedje JM (1999) 20 years since Dunedin: the past and future of microbial ecology. In Bell CR, Brylinksy M, Johnson-Green P (eds), *Microbial Biosystems: New Frontiers. Proceedings of the 8th International Symposium on Microbial Ecology* (unpaginated). Halifax, NS: Atlantic Canada Society for Microbial Ecology

Timmis JN, Ayliffe MA, Huang CY, Martin W (2004) Endosymbiotic gene transfer: organelle genomes forge eukaryotic chromosomes. *Nature Reviews Genetics* 5: 123–135

Treangen TJ, Rocha EPC (2011) Horizontal transfer, not duplication, drives the expansion of protein families in prokaryotes. *PLoS Genetics* 7(1): e1001284

True HL, Lindquist SL (2000) A yeast prion provides a mechanism for genetic variation and phenotypic diversity. *Nature* 407: 477–483

Turnbaugh PJ, Ley RE, Mahowald MA, Magrini V, Mardis ER, Gordon JI (2006) An obesity-associated gut microbiome with increased capacity for energy harvest. *Nature* 444: 1027–1031

Turnbaugh PJ, Hamady M, Yatsunenko T, Cantarel BL, Duncan A, Ley RE, Sogin ML, Jones WJ, Roe BA, Affourtit JP, et al. (2009) A core gut microbiome in obese and lean twins. *Nature* 457: 480–484

Turner D (2011) *Paleontology: A Philosophical Introduction.* Cambridge: Cambridge University Press

Valentine JW (1980) Determinants of diversity in higher taxonomic categories. *Paleobiology* 6: 444–450

van der Gast CJ, Jefferson B, Reid E, Robinson T, Bailey MJ, Judd SJ, Thompson IP (2005) Bacterial diversity is determined by volume in membrane bioreactors. *Environmental Microbiology* 8: 1048–1055

van Helvoort T (1994) The construction of bacteriophage as bacterial virus: linking endogenous and exogenous thought styles. *Journal of the History of Biology* 27: 91–139

van Iterson G Jr, den Dooren de Jong LE, Kluyver AJ (eds) (1983 [1940]) *Martinus Willem Beijerinck: His Life and His Work*. Madison, WI: Science Tech

van Niel CB (1949) The 'Delft School' and the rise of general microbiology. *Bacteriology Reviews* 13: 161–174

—(1955) Natural selection in the microbial world: the second Marjory Stephenson Memorial Lecture. *Journal of General Microbiology* 13: 201–217

—(1966) Microbiology and molecular biology. *Quarterly Review of Biology* 41: 105–112

van Ooij C (2011) First come, first served? *Nature Reviews Microbiology* 9: 698–699

van Regenmortel MHV (2008) The nature of viruses. In Mahy BWJ, van Regenmortel MHV (eds), *Encylopedia of Virology* (3rd ed) (pp. 19–23). Amsterdam: Elsevier

—(2011) Virus species. In Tibayrenc M (ed), *Genetics and Evolution of Infectious Diseases* (pp. 3–19). London: Elsevier

Vandamme P, Pot B, Gillis M, de Vos P, Kersters K, Swings J (1996) Polyphasic taxonomy, a consensus approach to bacterial systematics. *Microbiological Reviews* 60: 407–438

Varley JD, Scott PT (1998) Conservation of microbial diversity: a Yellowstone priority. *ASM News* 64: 147–151

Velicer GJ (2003) Social strife in the microbial world. *Trends in Microbiology* 11: 330–337

Velicer GJ, Vos M (2009) Sociobiology of the myxobacteria. *Annual Review of Microbiology* 63: 599–623

Venn AA, Loram JE, Douglas AE (2008) Photosynthetic symbioses in animals. *Journal of Experimental Botany* 59: 1069–1080

Verbruggen E, El Mouden C, Jansa J, Akkermans G, Bücking H, West SA, Kiers ET (2012) Spatial structure and interspecific cooperation: theory and an empirical test using the mycorrhizal mutualism. *American Naturalist* 179: E133–E146

Vergara-Silva F (2003) Plants and the conceptual articulation of evolutionary developmental biology. *Biology and Philosophy* 18: 249–284

Vijay-Kumar M, Aitken JD, Carvalho FA, Cullender TC, Mwangi S, Srinivasan S, Sitaraman SV, Knight R, Ley RE, Gerwirtz AT (2010) Metabolic syndrome and altered gut microbiota in mice lacking Toll-like receptor 5. *Science* 328: 228–231

Vos M (2009) Why bacteria engage in homologous recombination? *Trends in Microbiology* 17: 226–232

Waksman SA (1927) *Principles of Soil Microbiology*. London: Bailliere
—(1953) Streptomycin: background, isolation, properties, and utilization (Nobel Lecture, December 12th 1952). *Science* 118: 259–266

Walsh AM, Kortschak RD, Gardner MG, Bertozzi T, Adelson DL (2013) Widespread horizontal transfer of retrotransposons. *Proceedings of the National Academy of Sciences USA* 110: 1012–1016

Walter J, Ley R (2011) The human gut microbiome: ecology and recent evolutionary changes. *Annual Review of Microbiology* 65: 411–429

Ward BB (2005) Molecular approaches to marine microbial ecology and the marine nitrogen cycle. *Annual Review of Earth Planetary Sciences* 33: 301–333

Ward DM, Cohan FM, Bhaya D, Heidelberg JF, Kühl M, Grossman A (2008) Genomics, environmental genomics and the issue of microbial species. *Heredity* 100: 207–219

Williams TA, Foster PG, Cox CJ, Embey TM (2013) An archaeal origin of eukaryotes supports only two primary domains of life. *Nature* 504: 231–236

Waters CM, Bassler BL (2005) Quorum sensing: cell-to-cell communication in bacteria. *Annual Review of Cell and Developmental Biology* 21: 319–346

Weber B (2011) Life. In Zalta EN (ed), *Stanford Encyclopedia of Philosophy*, http://plato.stanford.edu/entries/life/

Weisberg M (2007) Three kinds of idealization. *Journal of Philosophy* 104: 639–659

Welch RA, Burland V, Plunkett G 3rd, Redford P, Roesch P, Rasko D, Buckles EL, Liou S-R, Boutin A, Hackett J., et al. (2002) Extensive mosaic structure revealed by the complete genome sequence of uropathogenic *Escherichia coli*. *Proceedings of the National Academy of Sciences USA* 99: 17020–17024

Weller JM (1955) Protista: non-plants, non-animals? *Journal of Paleontology* 29: 707–710

Wernergreen JJ (2012) Endosymbiosis. *Current Biology* 14: R555–R561

West SA, Kiers ET (2009) Evolution: what is an organism? *Current Biology* 19: R1080–R1082

West SA, Griffin AS, Gardner A (2007b) Social semantics: altruism, cooperation, mutualism, strong reciprocity and group selection. *Journal of Evolutionary Biology* 20: 415–432

West SA, Kiers ET, Simms EL, Denison RF (2002) Sanctions and mutualism stability: why do rhizobia fix nitrogen? *Proceedings of the Royal Society London B* 269: 685–694

West SA, Griffin AS, Gardner A, Diggle SP (2006) Social evolution theory for microorganisms. *Nature Reviews Microbiology* 4: 597–607

West SA, Diggle SP, Buckling A, Gardner A, Griffin AS (2007a) The social lives of microbes. *Annual Review of Ecology, Evolution, and Systematics* 38: 53–77

Wheelis ML, Kandler O, Woese CR (1992) On the nature of global classification. *Proceedings of the National Academy of Sciences USA* 89: 2930–2934

Whitaker RJ, Grogan DW, Taylor JW (2003) Geographic barriers isolate endemic populations of hyperthermophilic archaea. *Science* 301: 976–978

White JA (2011) Caught in the act: rapid, symbiont-driven evolution. *BioEssays* 33: 823–829

Whitfield JB, Asgari S (2003) Virus or not? Phylogenetics of polydnaviruses and their wasp carriers. *Journal of Insect Physiology* 49: 397–405

Whitman WB, Coleman DC, Wiebe WJ (1998) Prokaryotes: the unseen majority. *Proceedings of the National Academy of Sciences USA* 95: 6578–6583

—(2009) The modern concept of the prokaryote. *Journal of Bacteriology* 191: 2000–2005

Whittaker RH (1959) On the broad classification of organisms. *Quarterly Review of Biology* 34: 210–226

—(1969) New concepts of kingdoms and organisms. *Science* 163: 150–160

Wild G, Gardner A, West SA (2010) Wild, Gardner & West reply. *Nature* 463: E9

Wilkinson DM, Koumoutsaris S, Mitchell EAD, Bey I (2012) Modelling the effect of size on the aerial dispersal of microorganisms. *Journal of Biogeography* 39: 89–97

Williams GC (1966) *Adaptation and Natural Selection; A Critique of Some Current Evolutionary Thought*. Princeton, NJ: Princeton University Press

Williams TA, Embley TM, Heinz E (2011) Informational gene phylogenies do not support a fourth domain of life for nucleocytoplasmic large DNA viruses. *PLoS One* 6(6): e21080, doi:10.1371/journal.pone.0021080

Williams TA, Foster PG, Nye TMW, Cox CJ, Embley TM (2012) A congruent phylogenomic signal places eukaryotes within the Archaea. *Proceedings of the Royal Society London B* 279: 4870–4879

Williamson P, Wallace DWR, Law CS, Boyd PW, Collos Y, Croot P, Denman K, Riebesell U, Takeda S, Vivian C (2012) Ocean fertilization for geoengineering: a review of effectiveness, environmental impacts and emerging governance. *Process Safety and Environmental Protection* 90: 475–488

Wilson C (1995) *The Invisible World: Early Modern Philosophy and the Invention of the Microscope*. Princeton, NJ: Princeton University Press

Wilson EO (1992) *The Diversity of Life*. Cambridge, MA: Belknap/Harvard

—(1994) *Naturalist*. Washington, DC: Island Press

Winogradsky S (1949) *Microbiologie du Sol. Problèmes et Méthodes. Cinquante Ans de Recherches*. Paris: Masson et Cie

Wintermute EH, Silver PA (2010) Dynamics in the mixed microbial concourse. *Genes and Development* 24: 2603–2614

Woese CR (1996) Phylogenetic trees: whither microbiology? *Current Biology* 6: 1060–1063

—(1998) The universal ancestor. *Proceedings of the National Academy of Sciences USA* 95: 6854–6859

Woese CR, Fox GG (1977) Phylogenetic structure of the prokaryotic domain: the primary kingdoms. *Proceedings of the National Academy of Sciences USA* 74: 5088–5090

Woese CR, Magrum JL, Fox GE (1978) Archaebacteria. *Journal of Molecular Evolution* 11: 245–252

Woese CR, Kandler O, Wheelis ML (1990) Towards a natural system of organisms: proposal for the domains Archaea, Bacteria, and Eucarya. *Proceedings of the National Academy of Sciences USA* 87: 4576–4579

—(2006) How we do, don't and should look at bacteria and bacteriology. In Dworkin M, Falkow S, Rosenberg E, Schleifer K-H, Stackebrandt E (eds), *The Prokaryotes*, Vol 1 (pp. 3–23). Berlin: Springer-Verlag

Wolkowicz R, Schachter M (2008) What makes a virus a virus? *Nature Reviews Microbiology* (correspondence)

Wood EJF (1958) The significance of marine microbiology. *Bacteriological Reviews* 22: 1–19

—(1965) *Marine Microbial Ecology*. London: Chapman and Hall

—(1975) *The Living Ocean: Marine Microbiology*. London: Croom Helm

Woodcock S, Curtis TP, Head IM, Lunn M, Sloan WT (2006) Taxa-area relationships for microbes: the unsampled and the unseen. *Ecology Letters* 9: 805–812

Woodcock S, van der Gast CJ, Bell T, Lunn M, Curtis TP, Head IM, Sloan WT (2007) Neutral assembly of bacterial communities. *FEMS Microbiology Ecology* 62: 171–180

Xi Z, Bradley RK, Wurdack KJ, Wong KM, Sugumaran M, Bomblies K, Rest JS, Davis CC (2012) Horizontal transfer of expressed genes in a parasitic flowering plant. *BMC Genomics* 13: 227, doi:10.1186/1471-2164-13-227

Yoxen EJ (1979) Where does Schroedinger's 'What is Life' belong in the history of molecular biology? *History of Science* XVII: 17–52

Zaneveld JR, Nemergut DR, Knight R. 2008. Are all horizontal gene transfers created equal? Prospects for mechanism-based studies of HGT patterns. *Microbiology* 154: 1–15

Zarraonaindia I, Smith DP, Gilbert JA (2013) Beyond the genome: community-level analysis of the microbial world. *Biology and Philosophy* 28: 261–282

Zhang Q-G, Buckling A, Ellis RJ, Godfray HCJ (2009) Coevolution between co-operators and cheats in a microbial system. *Evolution* 63: 2248–2256

Zhang W-J, Chen C, Li Y, Song T, Wu L-F (2010) Configuration of redox gradient determines magnetotactic polarity of the marine bacteria MO-1. *Environmental Microbiology Reports* 2: 646–650

Zhaxybayeva O, Doolittle WF (2011) Lateral gene transfer. *Current Biology* 21: R242–R246

Zhu L, Wu Q, Dai J, Zhang S, Wei F (2011) Evidence of cellulose metabolism by the giant panda gut microbiome. *Proceedings of the National Academy of Sciences USA* 108: 17714–17719

Zinder ND, Lederberg J (1952) Genetic exchange in *Salmonella*. *Journal of Bacteriology* 64: 679–699

ZoBell CE (1946) *Marine Microbiology: a Monograph of Hydrobacteriology*. Walham MA: Chronica Botanica

ZoBell CE, Johnson FH (1949) The influence of hydrostatic pressure on the growth and viability of terrestrial and marine bacteria. *Journal of Bacteriology* 57: 179–189

Index